Informative Hypotheses

Theory and Practice for
Behavioral and Social Scientists

Chapman & Hall/CRC
Statistics in the Social and Behavioral Sciences Series

Series Editors

A. Colin Cameron
University of California, Davis, USA

J. Scott Long
Indiana University, USA

Andrew Gelman
Columbia University, USA

Sophia Rabe-Hesketh
University of California, Berkeley, USA

Anders Skrondal
Norwegian Institute of Public Health, Norway

Aims and scope

Large and complex datasets are becoming prevalent in the social and behavioral sciences and statistical methods are crucial for the analysis and interpretation of such data. This series aims to capture new developments in statistical methodology with particular relevance to applications in the social and behavioral sciences. It seeks to promote appropriate use of statistical, econometric and psychometric methods in these applied sciences by publishing a broad range of reference works, textbooks and handbooks.

The scope of the series is wide, including applications of statistical methodology in sociology, psychology, economics, education, marketing research, political science, criminology, public policy, demography, survey methodology and official statistics. The titles included in the series are designed to appeal to applied statisticians, as well as students, researchers and practitioners from the above disciplines. The inclusion of real examples and case studies is therefore essential.

Published Titles

Analysis of Multivariate Social Science Data, Second Edition
David J. Bartholomew, Fiona Steele, Irini Moustaki, and Jane I. Galbraith

Applied Survey Data Analysis
Steven G. Heeringa, Brady T. West, and Patricia A. Berglund

Bayesian Methods: A Social and Behavioral Sciences Approach, Second Edition
Jeff Gill

Foundations of Factor Analysis, Second Edition
Stanley A. Mulaik

Informative Hypotheses: Theory and Practice for Behavioral and Social Scientists
Herbert Hoijtink

Linear Causal Modeling with Structural Equations
Stanley A. Mulaik

Multiple Correspondence Analysis and Related Methods
Michael Greenacre and Jorg Blasius

Multivariable Modeling and Multivariate Analysis for the Behavioral Sciences
Brian S. Everitt

Statistical Test Theory for the Behavioral Sciences
Dato N. M. de Gruijter and Leo J. Th. van der Kamp

Chapman & Hall/CRC
Statistics in the Social and Behavioral Sciences Series

Informative Hypotheses
Theory and Practice for Behavioral and Social Scientists

Herbert Hoijtink

CRC Press
Taylor & Francis Group
Boca Raton London New York

CRC Press is an imprint of the
Taylor & Francis Group, an **informa** business
A CHAPMAN & HALL BOOK

CRC Press
Taylor & Francis Group
6000 Broken Sound Parkway NW, Suite 300
Boca Raton, FL 33487-2742

First issued in paperback 2019

ISBN-13: 978-1-4398-8051-7 (hbk)
ISBN-13: 978-0-367-38222-3 (pbk)

Library of Congress Cataloging-in-Publication Data

Hoijtink, Herbert.
    Informative hypotheses : theory and practice for behavioral and social scientists / Herbert Hoijtink.
       p. cm. -- (Chapman & Hall/CRC statistics in the social and behavioral sciences)
    Includes bibliographical references and index.
    ISBN 978-1-4398-8051-7 (hardback)
    1. Psychology--Statistical methods. 2. Social sciences--Statistical methods. 3. Hypothesis. I. Title.

BF39.H625 2011
300.72'7--dc23
                                                  2011039388

Visit the Taylor & Francis Web site at
http://www.taylorandfrancis.com

and the CRC Press Web site at
http://www.crcpress.com

To Anja, the love of my life.

Contents

5 Sample Size Determination: AN(C)OVA and Multiple Regression   77

Preface

Providing advice to behavioral and social scientists is the most interesting and challenging part of my work as a statistician. It is an opportunity to apply statistics in situations that usually have no resemblance to the clear-cut examples discussed in most textbooks on statistics. A fortiori, it is not unusual that scientists have questions to which I do not have a straightforward answer, either because the question has not yet been considered by statisticians, or because existing statistical theory cannot easily be applied because there is no software with which it can be implemented. An example of the latter are *informative hypotheses*. When I question scientists with respect to their theories, expectations, and hypotheses, they often respond with statements like, "I expect mean A to be bigger than means B and C"; "I expect that the relation between Y and both X1 and X2 is positive"; and "I expect the relation between Y and X1 to be stronger than the relation between Y and X2." Stated otherwise, they formulate their expectations in terms of inequality constraints among the parameters in which they are interested; that is, they formulate *informative hypotheses*.

In this book the evaluation of informative hypotheses is introduced for behavioral and social scientists. Chapters 1 and 2 introduce the univariate and multivariate normal linear models and the informative hypotheses that can be formulated in the context of these models. An accessible account of Bayesian evaluation of informative hypotheses is provided in Chapters 3 through 7. Chapter 8 provides an account of the non-Bayesian approaches for the evaluation of informative hypotheses for which software with which these approaches can be implemented is available. As is elaborated in Chapter 9, most of what is described in this book can be implemented using software that is freely available on the Internet: BIEMS and GenMVLData programmed by Joris Mulder, ConfirmatoryANOVA programmed by Rebecca Kuiper and Irene Klugkist, ContingencyTable programmed by Olav Laudy, GORIC programmed by Rebecca Kuiper, and Winbugs and Openbugs. One option is available in the commercial package SAS PROC PLM. In Chapter 10 the statistical foundations of Bayesian evaluation of informative hypotheses are presented. Associated with this book is a website containing software, manuals, the data and command files for the software examples presented in this book, and the unpublished papers that are referred to in this book. This website can be accessed via the home page of the author at http://tinyurl.com/hoijtink.

My research with respect to informative hypotheses started in 1994. At that time there were two books with respect to inequality constrained inference that summarized the state of the art: Barlow, Bartholomew, Bremner, and Brunk (1972) and Robertson, Wright, and Dykstra (1988). Both books focused on hypothesis testing for specific classes of statistical models, but software for the implementation of this approach was lacking. Visiting Don Rubin at Harvard's statistics department gave me the inspiration to explore the possibilities of the Bayesian approach. This resulted in the first paper on this topic together with Ivo Molenaar (Hoijtink and Molenaar, 1997). Around 2000, my project was reinforced with three PhD students: Irene Klugkist, Olav Laudy, and Bernet Kato. During their dissertation research we were visited by Antonio Forcina, who deepened our understanding of the evaluation of informative hypotheses using hypothesis testing. At the end of their dissertation research there was an international workshop in Utrecht that resulted in a special issue (number 59)

xiv

of *Statistica Neerlandica* in 2005. In 2005 I received a grant from the Netherlands Organization for Scientific Research (NWO-VICI-453-05-002) with which I could employ another five PhD students to work on informative hypotheses: Floryt van Wesel, Rebecca Kuiper, Carel Peeters, Rens van de Schoot, and Joris Mulder. These students were co-supervised by Irene Klugkist, Hennie Boeije, Jan-Willem Romeijn, Jean-Paul Fox, and Mervyn Silvapulle. The grant allowed me to invite Mervyn Silvapulle to Utrecht for a workshop on his book about inequality constrained inference (Silvapulle and Sen, 2004). It gave me the opportunity to organize another international workshop resulting in Hoijtink, Klugkist, and Boelen (2008). It finally allowed me to spend time abroad to find the peace to write this book. I am grateful to Iven van Mechelen and Katrijn van Deun (Catholic University Leuven) and Fred Rist (Westfalische Wilhelms Universitat), who provided stimulating writing environments. I am obliged to all these people. Without their contributions I could not have written this book.

Herbert Hoijtink, Utrecht, The Netherlands

Part I

Introduction

This part of the book contains Chapter 1 and Chapter 2. In Chapter 1 informative hypotheses are introduced in the context of the univariate normal linear model. Well-known instances of the univariate normal linear model are analysis of variance (ANOVA), analysis of covariance (ANCOVA), and multiple regression.

In Chapter 2 the multivariate normal linear model is used to further elaborate on the options that become available if informative hypotheses are used. Subsequently, the use of informative hypotheses in the context of multivariate one-sided testing, multivariate treatment evaluation, multivariate regression, and repeated measurement analysis is discussed.

Symbol Description

Chapter 1

μ_j	The mean in group j.
β_k	Regression coefficient relating the k-th predictor x to the dependent variable y.
σ^2	The residual variance.
ϵ_i	Residual for person i.
y_i	The score of person i on dependent variable y.
x_{ki}	The score of person i on predictor/covariate x_k.
d_{ji}	If the value is $1/0$, person i is/is not a member of group j.
$Z(y_i)$	Standardized value of y_i.
$Z(x_{ki})$	Standardized value of x_{ki}.
H_0	The null hypothesis.
H_a	The alternative hypothesis.
H_m	An informative hypothesis.
i	Index for person 1 to N.
d	Effect size used in the specification of an informative hypothesis.
j	Index for group 1 to J.

k	Index for predictor 1 to K.
m	Index for hypothesis 1 to M.

Chapter 2

y_{pi}	The score of person i on the p-th dependent variable y.
μ_{pj}	The mean in group j with respect to the p-th dependent variable y.
β_{pk}	Regression coefficient relating the k-th predictor x to the p-th dependent variable y.
ϵ_{pi}	Residual for person i on the p-th dependent variable y.
p	Index for dependent variable 1 to P.
$\boldsymbol{R}_m$	Matrix used to create inequality constraints for H_m.
$\boldsymbol{r}_m$	Vector used to include effect sizes in inequality constraints for H_m.
$\boldsymbol{S}_m$	Matrix used to create equality constraints for H_m.
$\boldsymbol{s}_m$	Vector used to include effect sizes in equality constraints for H_m.

Chapter 1

An Introduction to Informative Hypotheses

1.1 Introduction

Many researchers use analysis of variance (ANOVA; Rutherford, 2001; Weiss, 2006) when analyzing their data. Table 1.1 contains an example of a data matrix that can be analyzed with ANOVA. The table contains the scores on the dependent variable y for ten persons from four groups labeled 1 through 4. A typical analysis consists of two steps (let μ_j denote the population mean of group j for $j = 1, \ldots, 4$):

1. Test $H_0 : \mu_1 = \mu_2 = \mu_3 = \mu_4$, that is, all four means are equal ("nothing is going on"), against $H_a :$ not H_0, that is, not all four means are equal ("something is going on but I do not know what"). If this test favors H_a, continue with the second step.

2. Execute a pairwise comparison of means to determine which means are equal and which means are not equal.

If nothing more is known than that the dependent variable is named y and that there are four unspecified groups, such an exploratory approach is the only option. However, usually researchers have a clear idea about the meaning of y and the specification of the four groups. Suppose, for example, that y denotes a person's decrease in aggression level between week 1 (intake) and week 8 (end of training). Suppose furthermore, that forty persons in need of anger management training have randomly been assigned to one of four groups: 1, no training; 2, physical exercise; 3, behavioral therapy; and 4, physical exercise and behavioral therapy. In such a situation, researchers do not think "something is going on but I do not know what"; instead, they have clear ideas and expectations about what might be going on.

These expectations are called *informative hypotheses*. An example is $H_1 : \mu_1 < \{\mu_2 = \mu_3\} < \mu_4$ (note that $<$ means "smaller than"). This hypotheses states that the decrease

TABLE 1.1: Data Suited for ANOVA

Group 1 Nothing	Group 2 Physical	Group 3 Behavioral	Group 4 Both
1	1	4	7
0	0	7	2
0	0	1	3
1	2	4	1
−1	0	−1	6
−2	1	2	3
2	−1	5	7
−3	2	0	3
1	2	3	5
−1	1	6	4

in aggression level is smallest for the "no training" group, larger for the groups that receive physical exercise and behavioral therapy, respectively (with no preference for either method as expressed by the equality sign $=$), and largest for the group that receives both. A competing expectation is $H_2 : \{\mu_1 = \mu_2\} < \{\mu_3 = \mu_4\}$, that is, there is a positive effect of behavioral therapy, but not of physical exercise.

This book treats the specification and evaluation of informative hypotheses. In this chapter subsequently, ANOVA, analysis of covariance (ANCOVA, Howell, 2009, Chapter 16), and multiple regression (Kleinbaum, Kupper and Muller, 2008) are used to introduce and illustrate the specification of informative hypotheses. The chapter concludes with a section that briefly elaborates on the epistemological role of informative hypotheses.

1.2   Analysis of Variance

1.2.1   The Main Equation

As illustrated in the previous section, ANOVA can be used to investigate the relations among a set of $j = 1, \ldots, J$ means μ_j. ANOVA is a member of the family of univariate normal linear models (Cohen, 1968). These models and their multivariate extensions (see Chapter 2) provide the main context for the discussion of informative hypotheses in this book. The use of informative hypotheses in the context of other models is discussed in Chapter 7.

The ANOVA model is displayed in Figure 1.1, which will help us come to an understanding of the main equation of ANOVA. In an ANOVA, for the j-th group the data are assumed to have a normal distribution with mean μ_j and variance σ^2:

$$y_i = \mu_j + \epsilon_i, \tag{1.1}$$

where y_i for the 10 persons belonging to group j denotes the score of the i-th person on the dependent variable y (note that in the example presented in Table 1.1, each group contains 10 persons), ϵ_i is the residual, that is, the difference between the y_i and the corresponding mean; these residuals are assumed to have a normal distribution with mean 0 and variance σ^2, that is, $\epsilon_i \sim \mathcal{N}(0, \sigma^2)$.

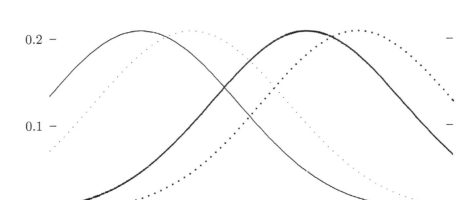

FIGURE 1.1: ANOVA model for the data in Table 1.1.

The curves in Figure 1.1 represent the distribution of y in each of the four groups according to the ANOVA model. The normality of the residuals is reflected by the shape of the curves. An important result for the ANOVA model is that the sample mean in each group is an estimate of the population mean. This is reflected by the location of the curves in Figure 1.1, which are equal to the sample means in Table 1.2. Another result is that the variance of each curve is equal to the pooled within variance across all groups. Because in the example the group sizes are equal, this is simply $(2.40 + 1.07 + 6.77 + 4.32)/4 = 3.64$.

Equation (1.1) can be adapted such that it applies to all groups simultaneously:

$$y_i = \mu_1 d_{1i} + ... + \mu_J d_{Ji} + \epsilon_i. \tag{1.2}$$

The main addition are J indicators of group membership: $d_{1i}, \ldots, d_{Ji}$. A person belonging to group j scores 1 on d_{ji} and 0 on the other indicators of group membership. Note that (1.2) with $d_{ji} = 1$ and the other indicators equal to 0 reduces to (1.1).

In this book we exemplify how informative hypotheses can be used to obtain information with respect to the population from which the data in Table 1.1 as displayed in Figure 1.1 are obtained. In the next sections, various options available when constructing informative hypotheses are presented and illustrated. The context of the illustrations will be the example with respect to reduction in aggression level.

TABLE 1.2: Sample Means and Variances for the Data in Table 1.1

	Group 1 Nothing	Group 2 Physical	Group 3 Behavioral	Group 4 Both
Sample Mean	−.20	.80	3.10	4.10
Variance	2.40	1.07	6.77	4.32

1.2.2   An Example Using Simple Constraints

Throughout the book the traditional null hypothesis is called H_0. For the aggression reduction example it is

$$H_0 : \mu_1 = \mu_2 = \mu_3 = \mu_4. \tag{1.3}$$

The null hypothesis is constructed using equality constraints denoted by "=". It is a very informative hypothesis because it states that all means are *exactly* equal to each other. This caused Cohen (1994) to call this hypothesis the "nill" hypothesis because he could not come up with research examples in which the null hypothesis might be a realistic representation of the population of interest. The point of view expressed by Cohen (1994) is further elaborated by Van de Schoot, Hoijtink, and Romeijn (2011). Royall (1997, pp. 79–81) is in line with Cohen and claims that the null hypothesis cannot be true, and consequently, that data are not needed in order to be able to reject it. Berger and Delampady (1987) are also critical with respect to the null hypothesis. However, they also show that inferences with respect to the null hypothesis are very similar to inferences with respect to hypotheses where strict equality constraints are replaced by a specific class (see Berger and Delampady (1987) for the criteria) of about equality constraints (see Section 1.2.3). Note furthermore that Wainer (1999) presents research examples where the null hypothesis might be of interest.

From now on the traditional alternative hypothesis will be denoted by H_a. It will be formulated as

$$H_a : \mu_1, \mu_2, \mu_3, \mu_4, \tag{1.4}$$

where a "," denotes that there are no constraints among the means adjacent to it. In words (1.4) states that the means are unconstrained. It is a very uninformative hypothesis: apparently something is going on, but it is unclear what is going on. Most researchers have clear ideas, expectations, or hopes with respect to what might be going on in their population of interest. Rejection of H_0 in favor of H_a does not provide those researchers with an evaluation of their "informative" ideas and expectations. In Section 1.2.4 a reformulation of the alternative hypothesis that is more interesting will be introduced.

Informative hypotheses will be denoted by H_m, for $m = 1, \ldots, M$, where M denotes the number of informative hypotheses under investigation. Two examples are

$$H_1 : \mu_1 < \{\mu_2 = \mu_3\} < \mu_4 \tag{1.5}$$

and

$$H_2 : \{\mu_1 = \mu_2\} < \{\mu_3 = \mu_4\}. \tag{1.6}$$

Three new elements appear in these hypotheses: "<" means smaller than, ">" means larger than, and "{...}" either contains a set of means that are constrained to be equal to each other, or, a set of unconstrained means. The constraints used are called "simple" because each specifies a relation between a pair of means, for example, $\mu_1 < \mu_2$ and $\mu_1 = \mu_2$. The interested reader is referred to Van Well, Kolk, and Klugkist (2008) and Van de Schoot, Hoijtink, and Doosje (2009) for an application of informative hypotheses constructed using simple constraints in psychological research. In Section 1.2.5, constraints among combinations of means are introduced.

1.2.3   Including the Size of an Effect

A classical application of null hypothesis significance testing may proceed along the following lines. Suppose that the null and alternative hypotheses for groups 1 and 4 from the example concerning reduction in aggression are

$$H_0 : \mu_1 = \mu_4, \tag{1.7}$$

and

$$H_a : \mu_1, \mu_4, \tag{1.8}$$

respectively. Let $\overline{y}_j$ denote the sample average of y in group j. Two questions must be answered: i) Is the difference between $\overline{y}_1$ and $\overline{y}_4$ significant or not? This can be done by means of the t-test for two independent groups, which is equivalent to an ANOVA with two groups. The larger the difference between the sample means of groups 1 and 4, the larger the probability that H_0 must be rejected in favor of H_a. The significance of a difference is indicated by the p-value computed for the t-test. A popular rule is to reject H_0 (a significant result) if the p-value is smaller than .05. For the data in Table 1.1 the p-value is .00; ii) If the difference is significant, the question arises whether or not the difference is relevant given the research question at hand. A popular effect size measure is Cohen's d (Cohen, 1992), which is given by

$$d = \frac{|\mu_1 - \mu_4|}{\sigma}, \tag{1.9}$$

where σ is the within group standard deviation, that is, the square root of the pooled within variance. According to Cohen, values of .8, .5, and .2 indicate large, medium, and small effects, respectively. An estimate of d is obtained using the sample means and standard deviation instead of μ_1, μ_4 and σ, respectively. For the example at hand $d = \frac{|-.20-4.10|}{\sqrt{3.36}} = -2.35$. Note that $3.36 = (2.40 + 4.32)/2$ (see Table 1.2).

Using informative hypotheses, both issues can be settled at once. Consider the following alternative for H_0:

$$H_1 : |\mu_1 - \mu_4| < d \times \sigma, \tag{1.10}$$

which can also be written as

$$H_1 : \begin{matrix} \mu_1 - \mu_4 > -d \times \sigma \\ \mu_1 - \mu_4 < d \times \sigma \end{matrix} ; \tag{1.11}$$

that is, the difference between the population means is smaller than d standard deviations. If H_1 is tested against H_a, it becomes clear at once whether the difference between the sample means is significant and relevant. The counterpart of equivalence testing (Wellek, 2003) is obtained if H_1 is tested against its complement H_{1_c} : not H_1. The goal of equivalence testing is to determine whether two means differ irrelevantly from each other or not. The comparison of a hypothesis with its complement is further elaborated in the next section.

The remaining question is how to specify $d \times \sigma$, which is similar to the second question from the previous paragraph "is the value of $d = -2.35$ indicative of a relevant difference or not?". There are two options: i) specify $d \times \sigma$ at once without bothering with d and σ separately, that is, which difference between two means is considered relevant; or ii) replace σ by the pooled within sample standard deviation and choose d such that it reflects a large, medium, or small effect according to Cohen.

Range constraints are a generalization of the about equality constraint presented in (1.10). Using an analogous representation as in (1.11), a range constraint can be formulated as:

$$H_1 : \begin{matrix} \mu_1 - \mu_4 > \eta_1 \\ \mu_1 - \mu_4 < \eta_2 \end{matrix}, \tag{1.12}$$

that is, the difference between both means is in the interval $[\eta_1, \eta_2]$ where η_1 denotes the lower bound and η_2 the upper bound of the range of interest. Using, for example, $\eta_1 = .2$ and $\eta_2 = .5$, a hypothesis is obtained that states that the expected effect size is (in terms of Cohen) between small and medium.

Effect size measures can also be included in informative hypotheses specified using inequality constraints. Consider the following two hypotheses that might be specified for the example concerning reduction in aggression level:

$$H_1 : \mu_1 < \mu_2 < \mu_3 < \mu_4, \qquad (1.13)$$

and

$$H_2 : \begin{array}{l} \mu_1 + d \times \sigma < \mu_2 \\ \mu_2 + d \times \sigma < \mu_3 \\ \mu_3 < \mu_4 \end{array} . \qquad (1.14)$$

Compared to H_1 in H_2, which is specified by means of three constraints, μ_2 is now at least $d \times \sigma$ larger than μ_1 (physical exercise leads to a reduction in aggression of at least d standard deviations). A similar statement is made for μ_2 and μ_3. Finally, it is specified that μ_4 is larger than μ_3 without including an effect size measure.

1.2.4   The Evaluation of One Informative Hypothesis

In classical statistical inference without informative hypotheses, there is a central role for the null hypothesis. For the example concerning reduction in aggression, the null hypothesis was

$$H_0 : \mu_1 = \mu_2 = \mu_3 = \mu_4. \qquad (1.15)$$

A problematic feature of the null hypothesis is that it is not often considered a realistic representation of the population under investigation. More often, researchers have a core idea, expectation, or research question that can be formulated as an informative hypothesis. For the example at hand, this informative hypothesis could be

$$H_1 : \mu_1 < \mu_2 < \mu_3 < \mu_4, \qquad (1.16)$$

which states that behavioral therapy leads to a larger reduction in aggression level than physical exercise, but that the combination of both approaches works best. If H_1 is the central research question, researchers should evaluate H_1 and not H_0.

The remaining question is how to evaluate H_1. Comparing it to H_0 does not make sense if H_0 is not considered a possible and realistic representation of the population of interest. However, there are two alternatives. First of all, H_1 can be compared with the alternative hypothesis H_a in which the means are unconstrained:

$$H_a : \mu_1, \mu_2, \mu_3, \mu_4. \qquad (1.17)$$

Note that H_1 can be seen as the combination of H_a and a set of constraints. The question that is answered if H_1 is compared to H_a is whether the set of constraints is supported by the data.

To explain the second option, consider the set of all models excluding H_1 that have a similar structure as H_1:

$$H_{1_c} : \begin{array}{c} \mu_1 < \mu_2 < \mu_4 < \mu_3 \\ \ldots \\ \mu_4 < \mu_3 < \mu_2 < \mu_1 \end{array} . \qquad (1.18)$$

The second option is to compare H_1 to H_{1_c} : not H_1, that is, are the means ordered as in H_1, or ordered differently? In statistical terms, H_{1_c} is called the complement (all other possibilities) of H_1. Note that in total there are $4! = 4 \times 3 \times 2 \times 1 = 24$ ways in which four means can be ordered.

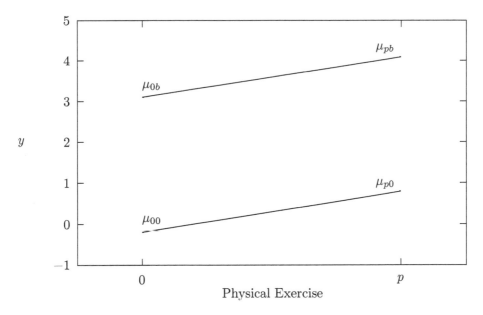

FIGURE 1.2: Representation of the means as a two way ANOVA.

1.2.5 Constraints on Combinations of Means

Consider again the example concerning reduction in aggression level. So far the four means have been treated individually. However, they can also be structured in two factors resulting in the layout of a two-way analysis of variance. The first factor indicates whether a person received physical exercise (p) or not (0); the second factor indicates whether a person received behavioral therapy (b) or not (0). This leads to the following relabeling of the means: $\mu_1 = \mu_{00}$, $\mu_2 = \mu_{p0}$, $\mu_3 = \mu_{0b}$, and $\mu_4 = \mu_{pb}$. The structure of the means in a two-way analysis of variance is displayed in Figure 1.2.

A classical evaluation of the two-way layout focuses on three hypotheses. First of all, is there a main effect of physical exercise? That is,

$$H_0 : (\mu_{00} + \mu_{0b})/2 = (\mu_{p0} + \mu_{pb})/2 \tag{1.19}$$

versus

$$H_a : (\mu_{00} + \mu_{0b})/2 \neq (\mu_{p0} + \mu_{pb})/2. \tag{1.20}$$

As can be seen in Figure 1.2, the average of the means for category "no physical exercise" $(\mu_{00} + \mu_{0b})/2$ is somewhat smaller than the average for category "physical exercise" $(\mu_{p0} + \mu_{pb})/2$. This suggests, that there is a small main effect of physical exercise. Second, is there a main effect of behavioral therapy? That is,

$$H_0 : (\mu_{00} + \mu_{p0})/2 = (\mu_{0b} + \mu_{pb})/2 \tag{1.21}$$

versus

$$H_a : (\mu_{00} + \mu_{p0})/2 \neq (\mu_{0b} + \mu_{pb})/2. \tag{1.22}$$

As can be seen in Figure 1.2, the average of the means for category "behavioral therapy" is larger than the average for category "no behavioral therapy." This suggests that there is a main effect of behavioral therapy. And finally, is there an interaction effect of physical exercise and behavioral therapy? That is,

$$H_0 : (\mu_{00} - \mu_{0b}) = (\mu_{p0} - \mu_{pb}) \tag{1.23}$$

versus

$$H_a : (\mu_{00} - \mu_{0b}) \neq (\mu_{p0} - \mu_{pb}). \tag{1.24}$$

As can be seen in Figure 1.2, the effect of behavioral therapy for the category "no physical exercise" $(\mu_{00} - \mu_{0b})$ is the same as the effect of behavioral therapy for the category "physical exercise." This implies that there is no interaction between physical exercise and behavioral therapy.

Here too, informative hypotheses can be formulated, but now they are specified using constraints on combinations of means. See Van Wesel, Klugkist, and Hoijtink (Unpublished) for a further elaboration of informative hypotheses in the context of the evaluation of interaction effects. A relevant informative hypothesis for the example at hand might be

$$H_2 : \begin{array}{c} (\mu_{00} + \mu_{0b})/2 < (\mu_{p0} + \mu_{pb})/2 \\ (\mu_{00} + \mu_{p0})/2 < (\mu_{0b} + \mu_{pb})/2 \\ (\mu_{0b} - \mu_{00}) < (\mu_{pb} - \mu_{p0}). \end{array} \tag{1.25}$$

This hypothesis states that there is a positive effect of physical exercise, a positive effect of behavioral therapy, and a stronger effect of behavioral therapy if it is combined with physical exercise. This hypothesis can be evaluated by comparing it to the combination of null hypotheses specified in (1.19), (1.21), and (1.23), that is, if no effect whatsoever of physical exercise and behavioral therapy is a possible and realistic state of affairs in the population of interest. To make a connection to the previous section, H_2 can also be evaluated by comparing it to the combination of alternative hypotheses in (1.20), (1.22), and (1.24), or to H_{2_c} : not H_2. Finally, it can also be compared to other informative hypotheses specified using constraints on (combinations) of means. An example is

$$H_2 : \begin{array}{c} (\mu_{00} + \mu_{0b})/2 = (\mu_{p0} + \mu_{pb})/2 \\ (\mu_{00} + \mu_{p0})/2 < (\mu_{0b} + \mu_{pb})/2 \\ (\mu_{00} - \mu_{0b}) = (\mu_{p0} - \mu_{pb}), \end{array} \tag{1.26}$$

that is, there is only an effect of behavioral therapy on the reduction in aggression level.

1.3   Analysis of Covariance

The analysis of covariance model (ANCOVA; Rutherford, 2001; Howell, 2009, Chapter 16) can be written as

$$y_i = \mu_1 d_{1i} + ... + \mu_J d_{Ji} + \beta_1 x_{1i} + ... + \beta_K x_{Ki} + \epsilon_i. \tag{1.27}$$

The main change with respect to the equation of an ANOVA (1.2) is the addition of $k = 1, ..., K$ continuous predictors that will be called covariates. The score of the i-th person on the k-th predictor is denoted by x_{ki}. The regression coefficient relating the k-th predictor to y is denoted by β_k.

Hypothetical data that will be used to illustrate ANCOVA are displayed in Table 1.3. The four groups and the dependent variable have the same meaning as in the ANOVA example. A covariate is available in the form of a person's age. Suppose the age of the

	TABLE 1.3: Data Suited for ANCOVA						
	Group 1 Nothing		Group 2 Physical		Group 3 Behavioral		Group 4 Both
	y	x	y	x	y	x	y	x
	0	18	3	23	4	21	6	21
	0	20	1	24	3	22	5	22
	0	21	1	19	4	23	5	23
	1	22	1	20	5	25	6	25
	2	23	2	21	5	26	6	24
	1	24	1	18	5	27	6	23
	0	19	1	20	4	23	6	26
	0	21	2	22	3	21	6	27
	0	20	3	23	5	22	6	24
	1	22	1	21	4	25	5	23
Mean	.50	21.00	1.60	21.10	4.20	23.50	5.70	23.80

persons is between 18 and 27. In such a situation it may very well be that younger persons must be assigned to groups 1 and 2 because they still have to mature and learn how to deal with their temper. Behavioral therapy might be considered too strong a measure for these persons. Similarly, it may very well be that older persons who have matured and nevertheless cannot control their temper should have behavioral therapy. For persons of average age, these arguments do not apply; they can be assigned to each of the four groups.

In this case a direct comparison of the groups by means of an ANOVA would not be fair because groups 1 and 2 score lower on the covariate than groups 3 and 4 (see the last line of Table 1.3 for the sample means). This difference might explain why the treatment effect is smaller for the first two groups. In such a nonrandomized experiment, the research questions and hypotheses should not be phrased in terms of means. An alternative is to phrase the hypotheses in terms of adjusted means, that is, means that account for the differences between the groups with respect to the scores on the covariate. As will be explained in the sequel, this can be achieved by means of an ANCOVA.

Figure 1.3 contains a visual representation of the data in Table 1.3 for the ANCOVA model with one covariate:

$$y_i = \mu_1 d_{1i} + ... + \mu_4 d_{4i} + \beta x_i + \epsilon_i. \tag{1.28}$$

The solid-line arrows pointing to the y-axis denote the location of the unadjusted means in each of the four groups. The dashed-line arrows denote the location of the adjusted means. As can be seen, the adjusted means are the predicted values of y for a person that scores 22.35 (the overall mean of x) on the covariate. As can be seen, the differences between the adjusted means are smaller than the differences between the unadjusted means. This implies that part of the differences between the unadjusted means are explained by the fact that the groups score differently on the covariate (indicated by the arrows pointing to the x-axis). The meaning of the regression coefficient β is indicated in the lower right-hand corner of Figure 1.3: if x increases by an amount of 1, y increases by an amount of β. For the data in Table 1.3, β equals .20. It is important to note that the μs in (1.28) only correspond to the adjusted means displayed in Figure 1.3 if the xs are standardized, that is, if each x_i is replaced by

$$Z(x_i) = \frac{x_i - \bar{x}}{SD_x}, \tag{1.29}$$

where $\bar{x}$ denotes the sample mean of x and SD_x the sample standard deviation of x. Without

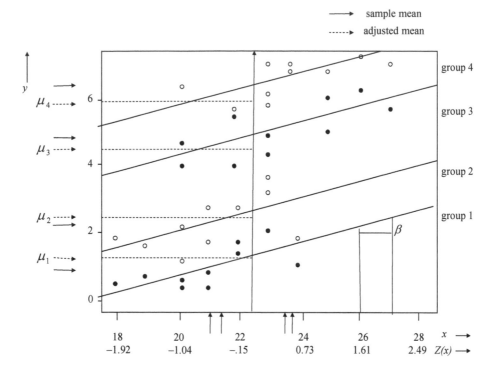

FIGURE 1.3: ANCOVA model for the data in Table 1.3.

standardization the μs are intercepts. Note furthermore that the software presented in Chapter 9 is calibrated properly only if the covariates are standardized. Rewriting (1.28) using a standardized covariate renders

$$y_i = \mu_1 d_{1i} + \ldots + \mu_4 d_{4i} + \beta Z(x_i) + \epsilon_i. \tag{1.30}$$

In Figure 1.3, standardized age $Z(x)$ can be found below the coding in years x.

For ANCOVA, informative hypotheses can be formulated in the same way as for ANOVA. The main difference is that hypotheses are formulated in terms of adjusted instead of unadjusted means to account for differences between the groups of interest with respect to one or more covariates. Stated otherwise, all the hypotheses formulated in Sections 1.2.2, 1.2.3, 1.2.4, and 1.2.5 have their counterparts in terms of adjusted means. In Section 2.7.2, a hypothesis addressing interaction between group and covariate is presented.

1.4   Multiple Regression

1.4.1   Introduction

In the classical formulation, the multiple regression model (Allison, 1999; Kleinbaum, Kupper, and Muller, 2008) contains only continuous predictors x:

$$y_i = \beta_0 + \beta_1 x_{1i} + \ldots + \beta_K x_{Ki} + \epsilon_i, \tag{1.31}$$

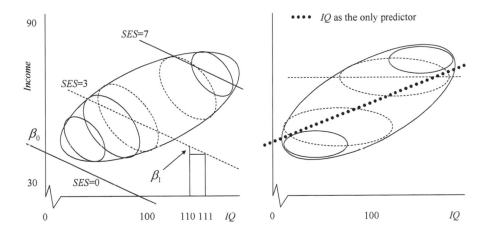

FIGURE 1.4: A visual illustration of regression models.

where β_0 is the intercept of the regression model, β_k for $k = 1, \ldots, K$ the unstandardized regression coefficient relating each predictor to y, and ϵ_i for $i = 1, \ldots, N$ the residuals in the prediction that, like for ANOVA and ANCOVA, are assumed to be normally distributed with mean 0 and variance σ^2, that is, $\epsilon_i \sim \mathcal{N}(0, \sigma^2)$.

Let y denote a person's income in units of 1,000 Euro per year, x_1 a person's intelligence quotient (IQ), and x_2 a person's social economic status (SES). For this situation, (1.31) reduces to

$$Income_i = \beta_0 + \beta_1 IQ_i + \beta_2 SES_i + \epsilon_i. \tag{1.32}$$

To illustrate the meaning of the intercept and the regression coefficients, (1.32) is displayed in Figure 1.4 for two hypothetical data sets. First the left-hand figure will be discussed. The distribution of the scores with respect to $Income$ and IQ is represented by the largest ellipse. As can be seen, there is a positive relationship between IQ and $Income$: the larger the scores on IQ, the larger the scores on $Income$. This feature however is *not* related to the model in (1.32) because in addition to IQ there is another predictor: SES. The smaller ellipses in Figure 1.4 represent the distribution of the scores with respect to $Income$ and IQ for each level of SES. The slopes of the solid and dashed lines show that for each level of SES there is a negative relationship between $Income$ and IQ. This conditional (on the level of SES) relationship between $Income$ and IQ is expressed by β_1 in (1.32). As is illustrated in the figure, β_1 is, for each level of SES, the decrease in $Income$ if IQ increases by one point (e.g., from 110 to 111).

The conditional (on the level of IQ) relationship between $Income$ and SES is expressed by β_2. This coefficient is not shown in Figure 1.4. This can only be done in a figure in which SES is placed on the x-axis, and the larger ellipse is divided into smaller ellipses using the levels of IQ. The intercept β_0 is illustrated in the figure; it is the point where the regression line intersects with the y-axis for $IQ = 0$ and $SES = 0$.

TABLE 1.4: Data Suited for Multiple Regression

Income	IQ	SES	Income	IQ	SES
51.09	79	1	66.12	97	4
40.88	82	3	66.81	100	5
41.29	83	4	69.09	100	4
57.74	85	4	80.32	100	5
57.01	87	4	64.47	110	4
41.18	88	3	65.20	116	6
58.24	95	2	80.31	117	7
52.19	95	3	83.93	120	5
59.54	96	5	67.84	121	4
67.40	96	3	64.46	123	5
Mean			61.76	99.50	4.05
SD			12.38	13.94	1.36

The right-hand figure in Figure 1.4 displays another hypothetical data set. Ignoring *SES*, that is, using a regression model with only *IQ* as the predictor,

$$Income_i = \beta_0 + \beta_1 IQ_i + \epsilon_i, \tag{1.33}$$

it can be seen (see the line consisting of dots) that there is a positive relationship between *IQ* and *Income*. Stated otherwise, β_1 from (1.33) is positive. However, if *SES* is added to the model as in (1.32), there is no relationship between *Income* and *IQ* for each level of *SES*. As is indicated by the dashed line, the regression coefficient β_1 from (1.32) is zero. What can be learned from this exposition is that the unconditional relationship between x_1 and y may be quite different from the relationship between x_1 and y conditional on all the other predictors in the regression equation.

A hypothetical data matrix that can be analyzed with (1.32) is displayed in Table 1.4. Interesting informative hypotheses for this data set could be

$$H_1 : \beta_1 > \beta_2, \tag{1.34}$$

that is, *IQ* is a stronger predictor than *SES*, and

$$H_2 : \beta_1 > 0, \beta_2 > 0, \tag{1.35}$$

that is, an increase in *IQ* and *SES* leads to an increase in *Income*. However, before evaluating informative hypotheses in the context of multiple regression, two issues must be considered: should the informative hypotheses be formulated in terms of unstandardized or standardized regression coefficients?; and, is the interest in the unconditional or conditional relation between each predictor and the dependent variable?

The importance of the first issue is illustrated in Table 1.5. The unstandardized regression coefficients show that an increase in *IQ* of 1 point leads to an increase in *Income* of .51,

TABLE 1.5: Unstandardized and Standardized Regression Coefficients

	Unstandardized	Standardized
β_0	2.19	61.76
β_1	.51	7.17
β_2	2.08	2.82

and an increase in SES of 1 level leads to an increase in $Income$ of 2.08. Does this imply that SES is a stronger predictor of $Income$ than IQ? Not necessarily. IQ and SES are measured on a different scale. IQ has a standard deviation of 13.94 and SES of 1.36. This implies that the relative size of a step of 1 on the SES scale is $13.94/1.36 = 10.25$ times as large as a step of 1 on the IQ scale. Stated otherwise, the unstandardized regression coefficients of IQ and SES not only indicate the strength of the relationship with $Income$, but also the scale of the predictor.

This problem is avoided if the predictors are standardized, that is,

$$Z(IQ_i) = (IQ_i - 99.50)/13.94$$
$$Z(SES_i) = (SES_i - 4.05)/1.36 \qquad (1.36)$$

rendering

$$Income_i = \beta_0 + \beta_1 Z(IQ_i) + \beta_2 Z(SES_i) + \epsilon_i. \qquad (1.37)$$

Both standardized predictors have the same scale, that is, a mean of 0 and a standard deviation of 1. The third column in Table 1.5 displays the regression coefficients obtained after standardization of IQ and SES. As can be seen, the intercept β_0 is now equal to the sample average of $Income$. Furthermore, the regression coefficient of $Z(IQ_i)$ is larger than the coefficient of $Z(SES_i)$, implying that the first is a stronger predictor than the latter. So unless a researcher has very good reasons not to standardize, evaluation of $H_1 : \beta_1 > \beta_2$ is only sensible if both predictors are standardized.

The second issue has more or less been treated during the discussion of Figure 1.4. As illustrated, there is a difference between the unconditional and the conditional relation between each predictor and the dependent variable. If a researcher wants to evaluate a hypothesis like $H_2 : \beta_1 > 0, \beta_2 > 0$, one of these perspectives must be chosen because it determines the results. Focussing on β_1 it can be seen both in the left- and right-hand figures of Figure 1.4 that the unconditional relationship of IQ with $Income$ is positive, while the conditional (on SES) relationship is negative and zero, respectively.

1.4.2 Informative Hypotheses

For the formulation of informative hypotheses in the context of multiple regression the same tools are available as for ANOVA. See Section 2.7.2 for an example of a hypothesis in the context of a quadratic regression model. However, the situation is more complicated if a researcher wants to include the size of an effect in an informative hypothesis. Consider the following alternative for $H_2 : \beta_1 > 0, \beta_2 > 0$:

$$H_3 : \beta_1 > d, \beta_2 > d, \qquad (1.38)$$

where d is the smallest effect size a researcher considers to be interesting. How can a researcher determine which values of d are small, medium, or large? An approach that works well is to standardize both the dependent variable and the predictors,

$$Z(Income_i) = \beta_0 + \beta_1 Z(IQ_i) + \beta_2 Z(SES_i) + \epsilon_i, \qquad (1.39)$$

and to determine the value of d for which the proportion of variance explained is at least R^2. Applying the formula

$$d = \sqrt{R^2/(K + (K^2 - K)\rho)} \qquad (1.40)$$

which is derived in Appendix 1A, to the example at hand renders $d = .31$. Note that the result is computed under the assumption that IQ and SES are predictors of equal strength, that the minimum proportion of variance explained that is interesting is at least $R^2 = .30$, that there are $K = 2$ predictors, and that the correlation between IQ and SES is $\rho = .6$. Appendix 1A also provides some further elaboration of the computation of effect sizes in the context of multiple regression models.

1.5   Epistemology and Overview of the Book

The previous sections contained an introduction to informative hypotheses for three statistical models that are often used for data analysis: ANOVA, ANCOVA, and multiple regression. Each of these models is a member of the univariate normal linear model. In the next chapter the multivariate normal linear model (Kim and Timm, 2007, Chapter 5) is introduced. Using a number of examples, the use of informative hypotheses to answer research questions formulated in the context of multivariate models is elaborated.

Often the following three components are needed for the evaluation of a research question: a population of interest, one or more hypotheses with respect to that population, and data sampled from that population. As may be clear by now, our focus will be on the hypotheses. However, first the population of interest and data sampled from that population are briefly discussed.

In the context of this book, a population is a constellation of persons and variables that provides the framework within which a research question can be formulated. To return to the first example from this chapter, the population consisted of persons with a rather low frustration tolerance threshold, and the variables of interest were reduction in the level of aggression and type of intervention. In the third example, the population consisted of mature adults, and the variables of interest were income, intelligence quotient, and social economic status. Note that the specification of a population in terms of persons and variables is not as easy as it seems in our simple examples. Sampling data from the population of interest is also not easy. Important issues are how to operationalize the variables of interest, for example, which instrument should be used to measure a person's level of aggression, and how persons should be randomized over treatment conditions. However, because this topic is beyond the scope of this book, the interested reader is referred to, for example, De Vaus (2001, 2002); Ryan (2007); and De Leeuw, Hox, and Dillman (2008) for a further discussion of population, measurement of variables, sampling, design, and many other issues that are related to the construction of a research project.

Our focus here is epistemological, that is, how to obtain knowledge with respect to the population of interest using data from that population to evaluate one or more hypotheses. This topic has been extensively researched and discussed by philosophers of science. Perhaps the most influential of them has been Popper (1959). Among other things, he discussed the specificity of hypotheses and falsification. According to Popper, hypotheses should be as specific as possible, because the more specific a hypothesis, the better it can be put to the test using the data. If the data are not in agreement with the implications of a hypothesis, the hypothesis is falsified; that is, it is no longer a candidate to describe the state of affairs in the population and must be replaced or adjusted.

Two other philosophers in this arena are Sober (2002) and Carnap (1952). Sober argues that a hypothesis should be plausible; that is, it should be parsimonious (or specific to use Popper's terminology) and supported by the data. In Carnap's view, one should not try to falsify a hypothesis but use the data to determine the degree of confirmation for a scientific hypothesis.

The combination of Popper's falsification principle and Sober's plausibility principle disqualifies both the traditional null and the traditional alternative hypothesis. The null is (almost always) to specific to be a realistic description of the population of interest. The alternative is the least specific and thus scientifically the least interesting hypothesis there is. Only very little is learned, if in a classical hypothesis test the null is rejected in favor of the alternative (Van de Schoot, Hoijtink, and Romeijn, 2011). How to formulate a falsifiable and plausible *informative hypothesis* is an interesting question. Three aspects can be distinguished in this question: Is the hypothesis scientifically interesting? How much is learned if the data strongly support the hypothesis? And, is there agreement among scientists how the hypothesis should be formulated to adequately represent an expectation with respect to the population of interest? The interested reader is referred to Van Wesel, Alisic, and Boeije (Unpublished) for an approach in which multiple sources of information are used to construct informative hypotheses.

As will be elaborated in Chapters 3 and 4, the Bayesian approach, and, as will be elaborated in Chapter 8, classical information criteria combine the complexity and fit of a hypothesis and use it to quantify the evidence in the data in favor of the hypotheses under investigation. This is very much in line with Sober plausibility principle and Carnap's "degree of confirmation." In Chapters 4 and 5, the application of informative hypotheses in the context of the univariate normal linear model is elaborated. Chapter 5 also introduces error probabilities (the counterpart of the errors of the first and second kind) and a simple procedure that can be used to obtain an indication of the sample sizes needed in order to control the error probabilities. In Chapter 6 the context is changed to the multivariate normal linear model. Among other things it contains a further elaboration of error probabilities and sample size determination. Chapter 7 provides an overview of Bayesian procedures for the evaluation of informative hypotheses in the context of contingency tables, multilevel models, latent class models, and statistical models in general. Null hypothesis significance testing is a procedure that can be used to falsify hypotheses. As may be clear by now, authors like Cohen (1994); Van de Schoot, Hoijtink, and Romeijn (2011); and Royall (1997) question the usefulness of the traditional null hypothesis H_0 : "nothing is going on." This is consistent with Sober's plausibility principle. However, as will be elaborated in Chapter 8, it is also possible to assign the role of null hypothesis to an informative hypothesis and subsequently evaluate it using null hypothesis significance testing. In Chapter 9 an overview is given of software that can be used for the evaluation of informative hypotheses using the Bayesian approach, an information criterion and null hypothesis significance testing. The book concludes with Chapter 10, which contains the statistical foundations of Bayesian evaluation of informative hypotheses.

1.6   Appendix 1A: Effect Sizes for Multiple Regression

The multiple regression model with standardized dependent en predictor variables is

$$Z(y_i) = \beta_0 + \sum_{k=1}^{K} \beta_k Z(x_{ki}) + \epsilon_i, \tag{1.41}$$

and the informative hypothesis of interest is

$$H_i : \beta_1 > d, \ldots, \beta_K > d. \tag{1.42}$$

The variance of the left-hand side of (1.41) is equal to the variance of the right-hand side of (1.41):

$$VAR[Z(y_i)] = VAR[\beta_0 + \sum_{k=1}^{K} \beta_k Z(x_{ki})] + VAR(\epsilon_i). \tag{1.43}$$

Because the variance of standardized variables equals 1.0, $R^2 = 1 - \frac{VAR(\epsilon_i)}{VAR[Z(y_i)]} = 1 - VAR(\epsilon_i)$. This can be simplified to

$$R^2 = \sum_{k=1}^{K} \beta_k^2 + 2 \sum_{k<k'} \beta_k \beta_{k'} \rho_{kk'}, \tag{1.44}$$

where $\rho_{kk'}$ denotes the correlation between x_{ki} and $x_{k'i}$. Using the assumption that each predictor is equally powerful and has coefficient d and the assumption that the correlation between each pair of predictors is equal and has value ρ, an equation is obtained that renders the value d as a function of the proportion of variance R^2 explained by the predictors and ρ:

$$R^2 = Kd^2 + (K^2 - K)d^2\rho, \tag{1.45}$$

that is,

$$d = \sqrt{R^2/(K + (K^2 - K)\rho)}. \tag{1.46}$$

Filling in the minimum proportion of variance explained that is interesting and a value for ρ the effect size d of interest to the researcher can straightforwardly be computed using the positive root of (1.46). Note that ρ can, for example, be determined via an inspection of the correlation matrix of the predictors, and use of, for example, the average of the correlations.

If either the assumption of predictors that are equally strong and/or the assumption of homogeneous between predictor correlations is inappropriate for the problem at hand, researchers can use (1.44) as the point of departure for derivations based on different assumptions analogous to the one presented in this section. For example, for the hypothesis:

$$H_4 : \beta_1 > \beta_2 + d, \tag{1.47}$$

that is, the coefficient of β_1 is at least d larger than the coefficient of β_2, the counterpart of (1.45) is

$$R^2 = (\beta_2 + d)^2 + \beta_2^2 + 2(\beta_2 + d)\beta_2\rho_{12}. \tag{1.48}$$

As can be seen, this equation can only be solved for d after a value for β_2 has been specified. Momentarily the best way to proceed is to specify different values for β_2 and to choose one of the corresponding d values.

As may have become clear after reading this appendix, the evaluation of informative hypotheses including effect sizes in the context of multiple regression, is a topic in need of further research. Although the procedure proposed has a clear motivation (the point of departure is a minimum proportion of variance explained), there may very well be better ideas for the specification of effect sizes than the one elaborated in this appendix.

Chapter 2

Multivariate Normal Linear Model

2.1 Introduction

In Chapter 1 examples of informative hypotheses formulated in the context of the univariate normal linear model were presented and discussed. The univariate normal linear model (Cohen, 1968) is

$$y_i = \mu_1 d_{1i} + ... + \mu_J d_{Ji} + \beta_1 x_{1i} + ... + \beta_K x_{Ki} + \epsilon_i, \qquad (2.1)$$

that is, there is one dependent variable y, persons (indexed by i) may belong to one of J groups, and there may be up to K variables x. Note that it is assumed that the residuals have a normal distribution with mean 0 and variance σ^2, that is, $\epsilon_i \sim \mathcal{N}(0, \sigma^2)$. Several specifications of (2.1) have been discussed in Chapter 1: the ANOVA model, that is, (2.1) without x variables; the ANCOVA model, that is, (2.1) in which the x variables have the role of covariates; and, multiple regression, that is, (2.1) in which $\mu_1 d_{1i} + ... + \mu_J d_{Ji}$ is replaced by the intercept β_0 and the x variables have the role of predictors. Furthermore, Chapter 1 presented and discussed several constraints that can be used to construct informative hypotheses: simple constraints like $\mu_1 > \mu_2$, constraints including the size of an effect like $|\mu_1 - \mu_2| < d \times \sigma$, and constraints on combinations of means like $(\mu_1 - \mu_2) > (\mu_3 - \mu_4)$.

In this chapter the multivariate normal linear model (Kim and Timm, 2007, Chapter 5) is introduced. The main difference compared to the univariate normal linear model is that there is more than one dependent variable y. Several applications of this model, like multivariate one-sided testing (Li, Gao and Huang, 2003), multivariate treatment evaluation, multivariate regression, and repeated measures analysis (Davis, 2002), will be discussed. Furthermore attention is given to three situations in which informative hypotheses play an important role. In the context of multivariate one-sided testing, attention is given to the situation in which the goal is to evaluate one informative hypothesis. In the context of multiple regression attention is given to the situation in which a researcher wants to build a hypothesis from a set of informative components. In this situation the point of departure is a reference hypothesis with no or relatively few constraints. Subsequently, each informative component is combined with the reference hypothesis and evaluated. Finally, the reference hypothesis is combined with the informative components that are worthwhile for a final evaluation. In the context of repeated measures analysis, the situation where two competing expectations have been translated into two competing hypotheses is illustrated.

2.2   Multivariate Normal Linear Model

2.2.1   Introduction

In this section the general form of the multivariate normal linear model is introduced. This general form may be more difficult to understand than the specifications (multivariate one-sided testing, multivariate treatment evaluation, multivariate regression and repeated measures analysis) that will be discussed in subsequent sections. It is very well possible to first read the next four sections, then return to this section and to end with the last section, which discusses less standard applications of the multivariate normal linear model.

The general form is

$$y_{1i} = \mu_{11}d_{1i} + \ldots + \mu_{1J}d_{Ji} + \beta_{11}x_{1i} + \ldots + \beta_{1K}x_{Ki} + \epsilon_{1i}$$

$$\ldots \tag{2.2}$$

$$y_{Pi} = \mu_{P1}d_{1i} + \ldots + \mu_{PJ}d_{Ji} + \beta_{P1}x_{1i} + \ldots + \beta_{PK}x_{Ki} + \epsilon_{Pi}.$$

There are P dependent variables denoted by y that are labeled $p = 1, \ldots, P$. There are N persons labeled $i = 1, \ldots, N$, and y_{pi} denotes the score of the i-th person on the p-th dependent variable. It is assumed that each dependent variable is continuous, that is, has at least an interval measurement level.

There are J groups that are labeled $j = 1, \ldots, J$. Each group has a (adjusted) mean on each dependent variable that is denoted by μ_{pj}, that is, the (adjusted) mean of the j-th group on the p-th dependent variable. As was elaborated in Section 1.3 concerning analysis of covariance, adjusted means account for possible differences in mean scores of the groups with respect to the x-variables, that is, adjusted means reflect the differences in scores on y for groups of persons that have the same scores on the x-variables. As was elaborated in Section 1.3, the μs in (2.2) can only be interpreted as adjusted means if the xs are standardized. The variable d_{ji} is scored 1 if person i is a member of group j and 0 otherwise.

There are K predictor variables that are labeled $k = 1, \ldots, K$. The score of the i-th person on the k-th predictor is denoted by x_{ki}. The relationship between x_{ki} and y_{pi} is denoted by β_{pk}. As was elaborated in Section 1.4 concerning multiple regression, if there is more than one x-variable, the βs represent the conditional (upon the other predictors) relation between the x-variable at hand and the y-variable at hand. Furthermore, it was elaborated that βs representing relations with the same dependent variable are only comparable if the corresponding predictors x are standardized. In the multivariate normal linear model, βs belonging to different dependent variables can also be compared. These βs are only comparable if both the dependent variables y and the predictors x are standardized.

Finally, it is assumed that

$$\begin{pmatrix} \epsilon_{1i} \\ \ldots \\ \epsilon_{Pi} \end{pmatrix} \sim \mathcal{N} \left(\begin{bmatrix} 0 \\ \ldots \\ 0 \end{bmatrix}, \begin{bmatrix} \sigma_1^2 & \ldots & \sigma_{1P} \\ \ldots & \ldots & \ldots \\ \sigma_{1P} & \ldots & \sigma_P^2 \end{bmatrix} \right). \tag{2.3}$$

This implies that for the p-th dependent variable, the residuals have a mean of 0 and a variance of σ_p^2. Furthermore, the residuals of dependent variables p and p' are not independent but have covariance $\sigma_{pp'}$.

2.2.2 General Form of the Constraints

Let $\boldsymbol{\theta} = [\mu_{11}, \ldots, \mu_{1J}, \beta_{11}, \ldots, \beta_{1K}, \ldots, \mu_{P1}, \ldots, \mu_{Pj}, \beta_{P1}, \ldots, \beta_{PK}]$, then the general form of the constraints that can be used within the multivariate normal linear model is

$$\boldsymbol{R}_m \boldsymbol{\theta} > \boldsymbol{r}_m, \tag{2.4}$$

in combination with

$$\boldsymbol{S}_m \boldsymbol{\theta} = \boldsymbol{s}_m. \tag{2.5}$$

The matrix $\boldsymbol{R}_m$ contains Q rows and $(J + K)P$ columns, where Q denotes the number of inequality constraints. Note that for each of the P dependent variables there are J means and K predictors. This explains why $\boldsymbol{R}_m$ contains $(J+K)P$ columns. The vector $\boldsymbol{r}_m$ contains Q elements, one for each constraint. The matrix $\boldsymbol{S}_m$ contains L rows and $(J + K)P$ columns, where L denotes the number of equality constraints. The vector $\boldsymbol{s}_m$ contains L elements. In the next section a specific application of (2.2), (2.4,) and (2.5) will be given: multivariate one-sided testing. In the subsequent sections, more examples of specific applications will follow.

The constraints in (2.4) and (2.5) are so-called linear inequality and equality constraints. In the context of normal linear models and most of the models discussed in Chapter 7, linear constraints are very useful for the formulation of informative hypotheses. However, in Section 7.2, the formulation of informative hypotheses for contingency tables is discussed. For contingency tables it will be shown that the formulation of informative hypotheses is facilitated if nonlinear constraints are used.

2.3 Multivariate One-Sided Testing

Consider the data in Table 2.1. They contain the improvement in depression, anxiety and social phobia of 20 persons after receiving cognitive therapy for their complaints. Note that the number 1 implies that, for example, depression was reduced by 1 unit (each of the three variables was measured on a scale running from 1 to 20 before and after receiving therapy for 12 weeks), and that the number -2 implies that, for example, depression increased by 2 units.

For this example, the multivariate normal linear model (2.2) reduces to

$$y_{1i} = \mu_1 + \epsilon_{1i}$$

$$\ldots \tag{2.6}$$

$$y_{3i} = \mu_3 + \epsilon_{3i}.$$

As can be seen, these equations contain the means μ_1, μ_2, and μ_3 of improvement with respect to depression, anxiety, and social phobia, respectively. The specification is completed using

$$\begin{pmatrix} \epsilon_{1i} \\ \ldots \\ \epsilon_{3i} \end{pmatrix} \sim \mathcal{N} \left(\begin{bmatrix} 0 \\ \ldots \\ 0 \end{bmatrix}, \begin{bmatrix} \sigma_1^2 & \ldots & \sigma_{13} \\ \ldots & \ldots & \ldots \\ \sigma_{13} & \ldots & \sigma_3^2 \end{bmatrix} \right), \tag{2.7}$$

TABLE 2.1: Example Data for the Therapy Group

Person	Depression	Anxiety	Social Phobia
1	−1	−2	−1
2	1	1	1
3	0	2	0
4	2	1	1
5	3	2	1
6	−1	−1	1
7	0	1	2
8	0	0	−1
9	2	1	2
10	1	0	0
11	−1	−1	1
12	1	1	1
13	0	1	0
14	2	1	2
15	3	2	1
16	−1	−1	0
17	0	1	2
18	0	0	−1
19	2	1	2
20	1	−1	1

which reflects that the residuals and thus the three variables of interest may be correlated.

In the context of this example, the main question is whether there has been a beneficial effect of therapy. This question can be formalized in the following informative hypothesis:

$$H_1: \begin{array}{l} \mu_1 > 0 \\ \mu_2 > 0 \\ \mu_3 > 0 \end{array} , \tag{2.8}$$

that is, therapy has had a positive effect on each of the three variables of interest. As can be seen, this is a multivariate one-sided hypothesis.

In the current example, the researcher has one informative hypothesis that he would like to evaluate. In order to be able to do so, a competitor for H_1 is needed. To continue the discussion started in Section 1.2.4, if a researcher wants to know if the constraints in H_1 are supported by the data, a natural competitor would be a hypothesis that allows all combinations of means possible except those covered by H_1, that is, H_{1_c} : not H_1 also called the complementary hypothesis. For the example at hand, this hypothesis implies that there has not been a beneficial effect of therapy on depression, anxiety, and social phobia. This approach is in line with the philosophy underlying this book: a researcher should specify his expectations and determine whether they are supported by the data or not.

A more traditional approach would be to determine whether H_1 receives stronger support from the data than the traditional null hypothesis (Li, Gao, and Huang, 2003):

$$H_0: \begin{array}{l} \mu_1 = 0 \\ \mu_2 = 0 \\ \mu_3 = 0 \end{array} , \tag{2.9}$$

that is, there is no effect of therapy on the three variables of interest. This corresponds to the traditional view underlying null hypothesis significance testing that the point of departure is that there is no effect of therapy until proven otherwise. Both approaches are further illustrated and evaluated in Chapter 6. In each situation where there is one informative hypothesis of interest, a researcher should decide which approach best suits his purposes. The decision should account for the fact that Cohen (1994), Royal (1997, pp. 79–81), and Van de Schoot, Hoijtink, and Romeijn (2011) consider the null hypothesis to be of little or no interest. The decision should also account for the fact that Wainer (1999) does see a use for the null hypothesis, and that Berger and Delampady (1987) show that the null hypothesis is closely related to a corresponding hypothesis formulated using about equality constraints.

Using $\boldsymbol{\theta} = [\mu_1, \mu_2, \mu_3]$ and (2.4), H_1 can be represented by $\boldsymbol{R}_m\boldsymbol{\theta} > \boldsymbol{r}_m$, where

$$\boldsymbol{R}_m = \begin{bmatrix} 1 & 0 & 0 \\ 0 & 1 & 0 \\ 0 & 0 & 1 \end{bmatrix} \tag{2.10}$$

and $\boldsymbol{r}_m = [0\ 0\ 0]$. Note that the first row of $\boldsymbol{R}_m$ combined with the first element of $\boldsymbol{r}_m$ represents the constraint $\mu_1 > 0$ and that the last row/last element represents the constraint $\mu_3 > 0$. Similarly, using (2.5), H_0 can be represented by $\boldsymbol{S}_m\boldsymbol{\theta} = \boldsymbol{s}_m$, where

$$\boldsymbol{S}_m = \begin{bmatrix} 1 & 0 & 0 \\ 0 & 1 & 0 \\ 0 & 0 & 1 \end{bmatrix} \tag{2.11}$$

and $\boldsymbol{s}_m = [0\ 0\ 0]$. This kind of notation will be useful later in the book. Therefore the examples presented in this chapter will be used to illustrate the use of $\boldsymbol{R}_m$, $\boldsymbol{S}_m$, $\boldsymbol{r}_m$, and $\boldsymbol{s}_m$.

2.4 Multivariate Treatment Evaluation

The hypothetical data in Table 2.1 come from 20 persons who were treated for their problems. In Table 2.2 the data for the control group, that is, 20 persons who were not treated for their problems are presented. The only difference between the persons in the therapy group and the control group is that the former were receiving therapy and the latter were placed on a waiting list.

For this example the multivariate normal linear model (2.2) reduces to

$$y_{1i} = \mu_{11}d_{1i} + \mu_{12}d_{2i} + \epsilon_{1i}$$

$$\cdots \tag{2.12}$$

$$y_{3i} = \mu_{31}d_{1i} + \mu_{32}d_{2i} + \epsilon_{3i}.$$

As can be seen, these equations contain the means μ_{11}, μ_{21}, and μ_{31} of each of the three variables of interest in the therapy group, and, the means μ_{12}, μ_{22}, and μ_{32} for the control group. Note once more that d_{1i} equals 1 if a person is a member of the therapy group and 0 otherwise. Similarly, d_{2i} equals 1 if a person is member of the control group and 0 otherwise.

In the context of this design with a therapy group and a control group, the main research question is whether the improvement with respect to the three variables of interest is larger

TABLE 2.2: Example Data for the Control Group

Person	Depression	Anxiety	Social Phobia
1	0	0	−1
2	−1	1	0
3	−1	−1	0
4	0	0	0
5	1	−1	0
6	−2	0	−1
7	1	1	0
8	0	−1	−1
9	0	1	0
10	0	−1	0
11	−1	−1	−1
12	−1	0	−1
13	1	0	0
14	0	0	0
15	−1	0	0
16	−1	0	0
17	0	−1	0
18	0	0	1
19	0	−1	−1
20	−1	−1	0

in the therapy group than in the control group. This research question leads to the following informative hypothesis:

$$H_1 : \quad \begin{matrix} \mu_{11} > \mu_{12} \\ \mu_{21} > \mu_{22} \\ \mu_{31} > \mu_{32} \end{matrix} , \tag{2.13}$$

that is, the means in the therapy group are larger than the corresponding means in the control group. A traditional evaluation of this informative hypothesis is obtained if it is compared to the null hypothesis:

$$H_0 : \quad \begin{matrix} \mu_{11} = \mu_{12} \\ \mu_{21} = \mu_{22} \\ \mu_{31} = \mu_{32} \end{matrix} . \tag{2.14}$$

An evaluation more in line with the philosophy underlying this book is to determine whether the data support the expectations formalized in H_1 or not, that is, compare H_1 with its complement H_{1_c}. Both approaches will be further illustrated and evaluated in Chapter 6.

Using $\boldsymbol{\theta} = [\mu_{11}, \mu_{12}, \mu_{21}, \mu_{22}, \mu_{31}, \mu_{32}]$ and (2.4), H_1 can be represented by $\boldsymbol{R}_m \boldsymbol{\theta} > \boldsymbol{r}_m$, where

$$\boldsymbol{R}_m = \begin{bmatrix} 1 & -1 & 0 & 0 & 0 & 0 \\ 0 & 0 & 1 & -1 & 0 & 0 \\ 0 & 0 & 0 & 0 & 1 & -1 \end{bmatrix}, \tag{2.15}$$

and $\boldsymbol{r}_m = [0 \ 0 \ 0]$. Note that the first row of $\boldsymbol{R}_m$ represents the constraint $\mu_{11} - \mu_{12} > 0$, which is equivalent to $\mu_{11} > \mu_{12}$, that is, the first element of H_1. Similarly, using (2.5), H_0 can be represented by $\boldsymbol{S}_m \boldsymbol{\theta} = \boldsymbol{s}_m$, where $\boldsymbol{s}_m = [0 \ 0 \ 0]$ and

$$\boldsymbol{S}_m = \begin{bmatrix} 1 & -1 & 0 & 0 & 0 & 0 \\ 0 & 0 & 1 & -1 & 0 & 0 \\ 0 & 0 & 0 & 0 & 1 & -1 \end{bmatrix}. \tag{2.16}$$

2.5 Multivariate Regression

The hypothetical example data in Appendix 2A contain the scores of 100 children on two dependent variables: Arithmetic Ability (A, in line with the tradition in the Netherlands scored on a scale from $1 =$ very poor to $10 =$ outstanding) and Language Skills (L with the same scoring as A). Furthermore, Appendix 2A contains four predictors: the scores of the fathers (FA and FL) and mothers (MA and ML) of the 100 children on A and L.

For this example the multivariate normal linear model (2.2) reduces to

$$A_i = \beta_{A0} + \beta_{A1}FA_i + \beta_{A2}FL_i + \beta_{A3}MA_i + \beta_{A4}ML_i + \epsilon_{Ai}$$

$$(2.17)$$

$$L_i = \beta_{L0} + \beta_{L1}FA_i + \beta_{L2}FL_i + \beta_{L3}MA_i + \beta_{L4}ML_i + \epsilon_{Li},$$

where the subscript A denotes that the corresponding β is from the equation where A is the dependent variable. Note that for the multivariate normal model without groups, $\mu_{p1}d_{1i}, \ldots, \mu_{pJ}d_{Ji}$ will be replaced by β_{p0}, that is, the intercept of the regression equation. As can be seen in Equation (1.31), and the left-hand panel of Figure 1.4, the same holds for univariate multiple regression.

It is important to note that the two dependent variables and the four predictors are not standardized, and consequently, that the βs appearing in (2.17) are not standardized. In Section 1.4, it was elaborated in the context of multiple regression why it is important to standardize both the dependent variable and the predictors. The main issue is that the comparison of unstandardized regression coefficients using $<$, $>$, and $=$ is meaningless because unstandardized coefficients do not only reflect the strength of the conditional association of the predictor at hand and the dependent variable, but also the scale on which both the predictor and the dependent variable are measured. The same holds in the context of multivariate regression: the formulation of informative hypotheses only makes sense if both the dependent variables and the predictors are standardized, and consequently, that the βs are standardized:

$$Z(A_i) = \beta_{A0} + \beta_{A1}Z(FA_i) + \beta_{A2}Z(FL_i) + \beta_{A3}Z(MA_i) + \beta_{A4}Z(ML_i) + \epsilon_{Ai}$$

$$(2.18)$$

$$Z(L_i) = \beta_{L0} + \beta_{L1}Z(FA_i) + \beta_{L2}Z(FL_i) + \beta_{L3}Z(MA_i) + \beta_{L4}Z(ML_i) + \epsilon_{Li},$$

where $Z(\cdot)$ denotes a standardized variable.

Using the regression coefficients of the multivariate regression model (2.18), expectations will be formulated for the hypothetical example data displayed in Appendix 2A. As will be shown below, these expectations consist of a series of informative components that will be synthesized into one informative hypothesis. This is different from the situation in the previous two sections where there was only on informative hypothesis of interest. How to deal with a series of informative components will be further elaborated here and in Chapter 6.

The first step when working with a series of informative components is the formulation of a reference hypothesis that will serve as the point of departure for the inferences to be made. One option is to use the unconstrained hypothesis (in the example at hand, a hypothesis without constraints on the regression coefficients) as the reference hypothesis. Another option is to construct a reference hypothesis using constraints, not because it is of interest to evaluate whether or not these constraints are supported by the data, but because it is known or dictated by common sense that these constraints hold. These constraints will be common to all hypotheses under investigation. They can be seen as a further specification of (2.18) and add to the specificity of the hypotheses that will be investigated, which according to Popper and Sober will increase the falsifiability of the hypotheses under investigation and increase their scientific value. Common sense dictates that the relation between the arithmetic ability of the father and the mother on the one hand and the arithmetic ability of the child on the other hand should be positive, that is, $\beta_{A1} > 0$ and $\beta_{A3} > 0$. Similarly, for language skills, $\beta_{L2} > 0$ and $\beta_{L4} > 0$.

In the second step, researchers may formulate a number of informative components, that is, a series of elaborations of the reference hypotheses that may or may not be true. To determine the support in the data for each informative component, each will be compared to the reference hypothesis. This will further be elaborated and illustrated in Chapter 6. In the third step, the informative components that are supported by the data are combined, and the support in the data for this combination is determined.

An informative component may consist of constraints on βs that belong to the same dependent variable. First of all, it is expected that the arithmetic skills of the parents have a stronger relation to the arithmetic skills of the child than the language skills of the parents, that is, $\beta_{A1} > \beta_{A2}$ and $\beta_{A3} > \beta_{A4}$. Furthermore, for the language skills of the child, it is similarly expected that $\beta_{L2} > \beta_{L1}$ and $\beta_{L4} > \beta_{L3}$.

Another component that will be used in the sequel imposes constraints on βs that belong to different dependent variables. First of all, it is expected that the arithmetic ability of the parents is a stronger predictor of arithmetic ability of the child than of language skills of the child, that is, $\beta_{A1} > \beta_{L1}$ and $\beta_{A3} > \beta_{L3}$. Similar expectations can be formulated with respect to the language skills of the parents, that is, $\beta_{L2} > \beta_{A2}$ and $\beta_{L4} > \beta_{A4}$.

A final component expresses the expectation that language skills of the parents are more important for the prediction of arithmetic skills of the child than arithmetic skills of the parents are for the prediction of language skills of the child, that is, $\beta_{A2} > \beta_{L1}$ and $\beta_{A4} > \beta_{L3}$.

Addition of each informative component to the reference hypothesis

$$H_1 : \begin{array}{l} \beta_{A1} > 0 \\ \beta_{A3} > 0 \\ \beta_{L2} > 0 \\ \beta_{L4} > 0 \end{array} \tag{2.19}$$

renders the following set of hypotheses that must be evaluated against the reference hypotheses:

$$H_{1\&2} : \begin{array}{l} H_1 \\ \beta_{A1} > \beta_{A2} \\ \beta_{A3} > \beta_{A4} \ ; \\ \beta_{L2} > \beta_{L1} \\ \beta_{L4} > \beta_{L3} \end{array} \tag{2.20}$$

$$H_{1\&3} : \begin{array}{l} H_1 \\ \beta_{A1} > \beta_{L1} \\ \beta_{A3} > \beta_{L3} \ ; \\ \beta_{L2} > \beta_{A2} \\ \beta_{L4} > \beta_{A4} \end{array} \tag{2.21}$$

and

$$H_{1\&4} : \begin{array}{l} H_1 \\ \beta_{A2} > \beta_{L1} \ . \\ \beta_{A4} > \beta_{L3} \end{array} \tag{2.22}$$

If, for example, $H_{1\&2}$ and $H_{1\&4}$ receive more support from the data than the reference hypothesis, the final evaluation consists of a comparison of $H_{1\&2\&4}$ with H_1.

Using $\boldsymbol{\theta} = [\beta_{A0}, \beta_{A1}, \beta_{A2}, \beta_{A3}, \beta_{A4}, \beta_{L0}, \beta_{L1}, \beta_{L2}, \beta_{L3}, \beta_{L4}]$ and (2.4), $H_{1\&2}$ can be represented by $\boldsymbol{R}_m \boldsymbol{\theta} > \boldsymbol{r}_m$, where

$$
\boldsymbol{R}_m = \begin{bmatrix}
0 & 1 & 0 & 0 & 0 & 0 & 0 & 0 & 0 & 0 \\
0 & 0 & 0 & 1 & 0 & 0 & 0 & 0 & 0 & 0 \\
0 & 0 & 0 & 0 & 0 & 0 & 0 & 1 & 0 & 0 \\
0 & 0 & 0 & 0 & 0 & 0 & 0 & 0 & 0 & 1 \\
0 & 1 & -1 & 0 & 0 & 0 & 0 & 0 & 0 & 0 \\
0 & 0 & 0 & 1 & -1 & 0 & 0 & 0 & 0 & 0 \\
0 & 0 & 0 & 0 & 0 & 0 & -1 & 1 & 0 & 0 \\
0 & 0 & 0 & 0 & 0 & 0 & 0 & 0 & -1 & 1
\end{bmatrix}, \tag{2.23}
$$

and $\boldsymbol{r}_m = [0\ 0\ 0\ 0\ 0\ 0\ 0\ 0]$. The first four lines of $\boldsymbol{R}_m$ represent the constraints in H_1, the last four lines the additional constraints in $H_{1\&2}$. The first line states that $\beta_{A1} > 0$ and the last line that $\beta_{L4} - \beta_{L3} > 0$, that is, $\beta_{L4} > \beta_{L3}$. Similar matrices can be constructed for the other informative components and the hypothesis involved in the final evaluation.

2.6   Repeated Measures Analysis

Informative hypotheses can also be used for the evaluation of expectations in the context of repeated measures analysis (Davis, 2002). For applications in psychological research where informative hypotheses are used in the context of repeated measures analysis, the interested reader is referred to Kammers, Mulder, de Vignemont, and Dijkerman (2009); Van de Schoot, Hoijtink, Mulder, van Aken, Orobio de Castro, Meeus, and Romeijn (2011); and Bullens, Klugkist, and Postma, A. (In Press).

The hypothetical data in Table 2.3 contain the scores of 20 girls and 20 boys on a depression questionnaire (scores ranging from $0 = $ low until $20 = $ high) at the ages of 8, 11, 14, and 17. For this example, the multivariate normal linear model (2.2) reduces to

$$
y_{8i} = \mu_{8g} d_{gi} + \mu_{8b} d_{bi} + \epsilon_{8i}
$$

$$
\cdots \tag{2.24}
$$

$$
y_{17i} = \mu_{17g} d_{gi} + \mu_{17b} d_{bi} + \epsilon_{17i},
$$

where y_{8i} denotes the depression score of person i at the age of 8 years, μ_{8b} denotes the mean of the depression scores of 8-year-old boys (b), and d_{gi} is a dummy variable that is scored 1 if person i is a girl (g) and 0 otherwise. As can be seen from (2.24), each gender group has a mean depression score at the ages of 8, 11, 14, and 17.

Components with the following structure are often convenient to formulate informative hypotheses in the context of repeated measures analysis:

- Simple contrasts like $\mu_{11g} < \mu_{14g}$.

- Within contrasts including effect sizes like $|\mu_{11b} - \mu_{14b}| < d \times \sigma_*$. Note that σ_* is the standard deviation of $y_{11b} - y_{14b}$ in the population. An estimate of this standard deviation can be obtained as follows: compute for each boy the difference between the score on y at age 11 and age 14, and subsequently compute the standard deviation of the difference.

TABLE 2.3: Example Data for Repeated Measures Analysis

Depression at Age 8	11	14	17	Gender	Depression at Age 8	11	14	17	Gender
8	10	12	8	g	16	4	15	16	b
11	8	5	13	g	4	20	4	9	b
15	22	16	19	g	2	5	9	5	b
6	12	13	12	g	8	19	1	13	b
9	14	5	10	g	12	7	3	6	b
14	12	5	5	g	12	17	22	8	b
3	11	19	6	g	13	15	17	9	b
12	8	11	5	g	6	4	7	11	b
5	14	8	6	g	0	6	1	3	b
8	0	7	11	g	15	4	11	6	b
5	12	14	11	g	13	3	12	12	b
10	12	11	8	g	7	12	8	2	b
19	9	11	8	g	4	7	9	5	b
9	16	17	1	g	13	12	4	9	b
13	19	10	13	g	9	12	3	8	b
15	19	13	12	g	11	8	19	16	b
20	16	16	12	g	7	8	7	10	b
16	3	6	12	g	7	11	13	10	b
13	13	13	16	g	9	12	12	7	b
5	13	9	10	g	9	18	11	7	b

- Between contrasts including effect sizes like $|\mu_{11b} - \mu_{11g}| < d \times \sigma$. Note that σ^2 is the population pooled within group variance of y. An estimate of this variance an be obtained using the pooled within group variance in the sample.

- Within group difference contrasts like $\mu_{11g} - \mu_{8g} < \mu_{14g} - \mu_{11g}$, that is, the increase in depression for girls is larger between the ages of 11 and 14 than between the ages of 8 and 11.

- Between group difference contrasts like $\mu_{17g} - \mu_{14g} > \mu_{17b} - \mu_{14b}$, that is, the increase in depression for girls between the ages of 14 and 17 is larger than the corresponding increase in depression for boys. This contrast can equivalently be formulated as $\mu_{17g} - \mu_{17b} > \mu_{14g} - \mu_{14b}$.

- Combinations of means such as $-1.5\mu_{8g} - .5\mu_{11g} + .5\mu_{14g} + 1.5\mu_{17g} > -1.5\mu_{8b} - .5\mu_{11b} + .5\mu_{14b} + 1.5\mu_{17b}$. What is given before and after the $>$ is the linear contrast over time (see, for example, Davis, 2002, p. 53). If, for example, the mean depression scores of the girls are increasing over time (e.g. 4, 8, 12, and 16) the linear contrast has the value $-1.5 \times 4 - .5 \times 8 + .5 \times 12 + 1.5 \times 16 = 20$. Stated otherwise, if the linear contrast is positive as in the example given, this shows that the means are increasing over time, if the linear contrast is negative this shows that the means are decreasing. The interested reader is referred to Rosenthal, Rosnow, and Rubin (2000) for a comprehensive and accessible treatment of the use and evaluation of contrasts, including and beyond the linear contrast.

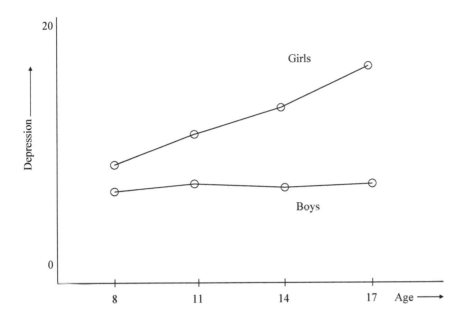

FIGURE 2.1: Visualization of H_1 for the repeated measures data.

For this example, two competing expectations with respect to the development of depression of girls and boys will be translated into informative hypotheses. The first expectation consists of three components. The first component states that over time the level of depression of the girls increases (see, the top line in Figure 2.1). The second component states that over time the level of depression of the boys is more or less constant (see the bottom line in Figure 2.1). The third component states that at each time point the girls are more depressed than the boys (see the difference between both lines in Figure 2.1). These expectations lead to the following informative hypothesis:

$$H_1: \quad \begin{array}{c} \mu_{8g} < \mu_{11g} < \mu_{14g} < \mu_{17g} \\ |\mu_{8b} - \mu_{11b}| < .7 \\ |\mu_{11b} - \mu_{14b}| < .7 \\ |\mu_{14b} - \mu_{17b}| < .7 \\ \mu_{8g} > \mu_{8b} \\ \mu_{11g} > \mu_{11b} \\ \mu_{14g} > \mu_{14b} \\ \mu_{17g} > \mu_{17b} \end{array} \quad . \tag{2.25}$$

Note that the standard deviations of $y_{8b} - y_{11b}$, $y_{11b} - y_{14b}$, and $y_{14b} - y_{17b}$ are all approximately 7. Therefore $d \times \sigma_* = .1 \times 7 = .7$, that is, the difference in the scores of the boys over time is smaller than .7 within standard deviations.

The second expectation also consists of three components. The first component states that at age 8 girls are more depressed than boys. The second component states that with increasing age the difference in depression between girls and boys is decreasing (see the difference between the two lines in Figure 2.2). The third component states that over time the level of depression of the boys is more or less constant. This leads to the following

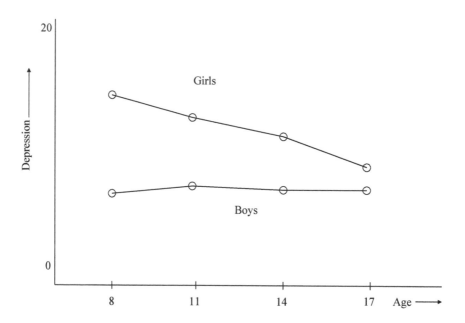

FIGURE 2.2: Visualization of H_2 for the repeated measures data.

informative hypothesis:

$$H_2 : \begin{array}{c} \mu_{8g} > \mu_{8b} \\ \mu_{8g} - \mu_{8b} > \mu_{11g} - \mu_{11b} > \mu_{14g} - \mu_{14b} > \mu_{17g} - \mu_{17b} \\ |\mu_{8b} - \mu_{11b}| < .7 \\ |\mu_{11b} - \mu_{14b}| < .7 \\ |\mu_{14b} - \mu_{17b}| < .7 \end{array} \quad . \tag{2.26}$$

What is rendered by the Bayesian approach for the evaluation of informative hypotheses (see Chapters 3 and 4) when two competing hypotheses are evaluated is the relative support in the data for both hypotheses, that is, which hypothesis is supported most, and to what degree is it supported more than the other? The implication is that the evaluation of a set of competing hypotheses may result in selection of the best of two inferior hypotheses. Note that this problem is not an issue if a hypothesis is compared to its complement, or, when an informative hypothesis is being built using informative components as was elaborated in the previous section. The problem can be avoided via the inclusion of the unconstrained hypothesis H_a as a fail-safe hypothesis in the set of hypotheses that is being evaluated. If the support in the data is larger for H_a than for a competing informative hypothesis H_m, then the constraints used are not supported by the data, and even is H_m is the best of a set of competing hypotheses under investigation, it is very likely that H_m does not provide an accurate description of the population of interest. This is further elaborated in Section 6.6.

Using $\boldsymbol{\theta} = [\mu_{8g}, \mu_{8b}, \mu_{11g}, \mu_{11b}, \mu_{14g}, \mu_{14b}, \mu_{17g}, \mu_{17b}]$ and (2.4), H_2 can be represented by $\boldsymbol{R}_m\boldsymbol{\theta} > \boldsymbol{r}_m$, where

$$
\boldsymbol{R}_m = \begin{bmatrix}
1 & -1 & 0 & 0 & 0 & 0 & 0 & 0 \\
1 & -1 & -1 & 1 & 0 & 0 & 0 & 0 \\
0 & 0 & 1 & -1 & -1 & 1 & 0 & 0 \\
0 & 0 & 0 & 0 & 1 & -1 & -1 & 1 \\
0 & 1 & 0 & -1 & 0 & 0 & 0 & 0 \\
0 & -1 & 0 & 1 & 0 & 0 & 0 & 0 \\
0 & 0 & 0 & 1 & 0 & -1 & 0 & 0 \\
0 & 0 & 0 & -1 & 0 & 1 & 0 & 0 \\
0 & 0 & 0 & 0 & 0 & 1 & 0 & -1 \\
0 & 0 & 0 & 0 & 0 & -1 & 0 & 1
\end{bmatrix}, \tag{2.27}
$$

and $\boldsymbol{r}_m = [0\ 0\ 0\ 0\ -.7\ -.7\ -.7\ -.7\ -.7\ -.7]$. Note that, $|\mu_{14b} - \mu_{17b}| < .7$ is represented by the last two lines in $\boldsymbol{R}_m$ and the last two numbers in $\boldsymbol{r}_m$ that state that $\mu_{14b} - \mu_{17b} > -.7$ and $\mu_{17b} - \mu_{14b} > -.7$.

2.7 Other Options

In the previous sections, several examples of applications of informative hypotheses in the context of the multivariate normal linear model have been given. Furthermore, three approaches that use informative hypotheses have been highlighted: evaluation of one informative hypothesis; evaluation of a series of nested informative hypotheses; and evaluation of competing informative hypotheses with the inclusion of a fail-safe hypothesis to avoid selection of the best of two inferior hypotheses. The examples given are by no means a complete overview of the options that are available in the context of the multivariate normal linear model. In this section a few other options are briefly highlighted.

2.7.1 Covariates

Most of the examples given in this chapter can be extended via the addition of covariates. Two examples will be given: the addition of a covariate to multivariate treatment evaluation as discussed in Section 2.4 and the addition of a (possibly time varying) covariate to the repeated measures analysis in Section 2.6.

The model in (2.12) is essentially the combination of three ANOVAs (see Section 1.2). In an ANOVA it may be useful to add a covariate to the model (see Section 1.3) to account for differences between the groups with respect to the covariate (Howell, 2009, Chapter 16). The same can be done in the context of multivariate treatment evaluation. Suppose that in addition to the variables in Tables 2.1 and 2.2, also the intelligence quotient (IQ) of each person is available (persons with a higher IQ may benefit more from cognitive therapy). Then standardized IQ can be added to (2.12) in order to account for possible differences between the therapy and control groups. This would change (2.12) to

$$
y_{1i} = \mu_{11}d_{1i} + \mu_{12}d_{2i} + \beta_1 Z(IQ_i) + \epsilon_{1i}
$$

$$
\cdots \tag{2.28}
$$

$$
y_{3i} = \mu_{31}d_{1i} + \mu_{32}d_{2i} + \beta_3 Z(IQ_i) + \epsilon_{3i},
$$

where β_3 is the regression coefficient relating $Z(IQ)$ to the third dependent variable. If (2.28) is used, the resulting conclusions with respect to the informative hypotheses (2.13) would *not* be in terms of means, but in terms of adjusted means. As can, for example, be seen in Figure 1.3, the adjusted mean of a group is the altitude of the regression line evaluated at the sample average of the covariate. Therefore, in this book and the accompanying software (see Chapter 9), it assumed that each covariate is standardized to have a mean of 0 and a variance of 1. Only for standardized covariates, the μs in (2.28) represent adjusted means; in all other cases the μs are intercepts.

The model in (2.24) has the same form as the model in (2.12), the main difference being that in the former the dependent variables are repeated measures (Davis, 2002) of the same variable (depression) while in the latter there are three different dependent variables (depression, anxiety, and social phobia). Addition of a standardized covariate such as $Z(IQ)$ measured at age 8 to model (2.24) renders

$$y_{8i} = \mu_{8b}d_{bi} + \mu_{8g}d_{gi} + \beta_8 Z(IQ_i) + \epsilon_{8i}$$

$$\dots \tag{2.29}$$

$$y_{17i} = \mu_{17b}d_{bi} + \mu_{17g}d_{gi} + \beta_{17} Z(IQ_i) + \epsilon_{17i},$$

where β_{17} relates $Z(IQ)$ to depression at the age of 17. The effect of adding $Z(IQ)$ to the repeated measures model is that the informative hypotheses (2.25) and (2.26) are now formulated in terms of means adjusted for group differences with respect to the covariate IQ. Note that the model could further be specified using the constraint $\beta_8 = \beta_{11} = \beta_{14} = \beta_{17}$, that is, the effect of IQ on depression is the same at each age. The interested reader is referred to Mulder, Klugkist, van de Schoot, Meeus, Selfhout, and Hoijtink (2009) for further elaboration and examples of the use of covariates in the context of the evaluation of informative hypotheses for repeated measures models. Here too it holds that the μs only represent adjusted means if the covariates are standardized.

It is also possible to add time-varying covariates to the repeated measures model (2.24). Suppose, for example, that for each child at each age not only the level of depression is recorded, but also a score representing their achievements (A) in school in the form of a grade on a scale (commonly used in the Netherlands) ranging from 1 = very poor to 10 = outstanding. Addition of the standardized covariate $Z(A)$ to (2.24) renders

$$y_{8i} = \mu_{8b}d_{bi} + \mu_{8g}d_{gi} + \beta_8 Z(A_{8i}) + \epsilon_{8i}$$

$$\dots \tag{2.30}$$

$$y_{17i} = \mu_{17b}d_{bi} + \mu_{17g}d_{gi} + \beta_{17} Z(A_{17i}) + \epsilon_{17i},$$

where β_{17} represents the relation between the covariate measured at age 17 and the depression level at age 17. The effect of adding $Z(A)$ to the repeated measures model is that the informative hypotheses (2.25) and (2.26) are now formulated in terms of means adjusted for group differences with respect to the covariate A measured at the age at which the level of depression is also measured. Note that the model could further be specified using the constraint $\beta_8 = \beta_{11} = \beta_{14} = \beta_{17}$, that is, the effect of A on depression is the same at each age.

As before, the μs in (2.30) can only be interpreted as adjusted means if the covariates are standardized. Note that standardization of time-varying covariates only makes sense if the covariates are jointly standardized, that is, each score on each covariate must be standardized using the average and standard deviation of the scores of all N persons on all P covariates. Consequently, the adjusted means are altitudes of regression lines evaluated not with respect to the average of the covariate for the time point at hand, but with respect to the overall average of all $N \times P$ times a covariate is scored.

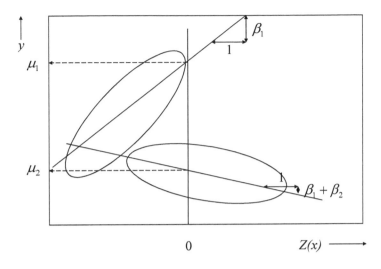

FIGURE 2.3: Interaction in an ANCOVA model.

2.7.2 Interactions and Quadratic Effects

Using specific applications of the univariate normal linear model (2.1), two more applications of informative hypotheses will be highlighted. The first concerns the specification of an interaction effect in an ANCOVA model (Allen, 1997, Chapter 32). The second concerns the specification of a quadratic regression (see Section 1.4 for an introduction to multiple regression and Allen, 1997, Chapter 26, for quadratic regression).

Consider a model with one dependent variable, two groups, and one covariate (standardized to have mean 0 and variance 1):

$$y_i = \mu_1 d_{1i} + \mu_2 d_{2i} + \beta_1 Z(x_i) + \beta_2 d_{2i} Z(x_i) + \epsilon_i. \tag{2.31}$$

A visualization of (2.31) in which the meaning of the μs and βs is highlighted can be found in Figure 2.3. As can be seen, there are two groups. The adjusted means with respect to y are denoted by μ_1 and μ_2. As can be seen, the relation between y and $Z(x)$ differs between both groups. The slope in group 1 is denoted by β_1 and is positive. The slope in group 2 is denoted by $\beta_1 + \beta_2$ (note that β_2 is the difference in slopes between group 1 and 2 because $d_{2i} = 1$ if a person is a member of group 2 and 0 otherwise). Note also that β_2 is negative.

Sometimes researchers encounter a situation as displayed in Figure 2.3 when they want to execute an ANCOVA, but an important ANCOVA assumption, that is, parallel regression lines, is violated (Allen, 1997, Chapter 32). An alternative for the now inappropriate ANCOVA is to formulate an informative hypothesis with respect to (2.31) that contains three components; for example,

$$H_1 : \begin{array}{c} \mu_1 > \mu_2 \\ \beta_1 > 0 \\ \beta_1 + \beta_2 < 0 \end{array} , \tag{2.32}$$

that is, the adjusted mean of group 1 is larger than the adjusted mean of group 2, the

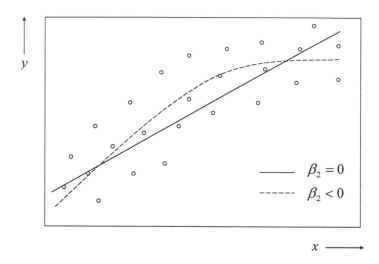

FIGURE 2.4: Quadratic regression.

slope of the regression line of group 1 is positive, and the slope of the regression line of group two is smaller than the slope of the regression line of group 1. The precise meaning of this informative hypothesis depends on the average score of each group with respect to the covariate. This should be taken into account while formulating the informative hypothesis. In combination with (2.31), H_1 addresses research questions with respect to adjusted means without the requirement of parallel regression lines. If the data look like the hypothetical example in Figure 2.3, the support for H_1 compared to H_{1_c} will be substantial.

Finally, consider the following regression model

$$y_i = \beta_0 + \beta_1 x_i + \beta_2 x_i^2 + \epsilon_i, \tag{2.33}$$

where β_1 denotes the linear effect of x on y, and β_2 the quadratic effect (Allen, 1997, Chapter 26). Example data that can be analyzed with (2.33) are presented in Figure 2.4. For these data, the question could be whether there is a ceiling effect in the relation between x and y (the dashed line) or not (the solid line). This question can be evaluated using the following informative hypotheses:

$$H_1 : \begin{array}{l} \beta_1 > 0 \\ \beta_2 = 0 \end{array}, \tag{2.34}$$

that is, there is a positive relationship between x and y, and there is no ceiling effect, and

$$H_2 : \begin{array}{l} \beta_1 > 0 \\ \beta_2 < 0 \end{array}, \tag{2.35}$$

that is, there is a positive relationship between x and y, and there is a ceiling effect. Note that in this specific example, standardization of y and/or x and x^2 is not necessary because there is no mutual comparison of β_1 and β_2 in the informative hypotheses.

2.8 Appendix 2A: Example Data for Multivariate Regression

Child	A	L	FA	FL	MA	ML
1	7	6	2	5	4	4
2	3	5	5	6	6	5
3	4	4	6	7	6	7
4	5	3	7	5	4	5
5	5	6	3	3	3	2
6	4	5	5	5	4	4
7	7	8	8	6	8	8
8	3	4	7	6	6	7
9	4	4	7	7	4	7
10	3	6	3	6	5	7
11	6	6	5	4	5	4
12	4	6	5	5	6	4
13	4	1	5	4	6	4
14	2	4	5	5	5	6
15	4	4	3	3	4	7
16	3	8	5	6	7	8
17	5	3	5	5	6	2
18	5	8	5	6	5	6
19	5	5	4	4	6	7
20	7	4	2	5	5	1
21	4	1	5	2	5	4
22	5	4	8	4	5	5
23	5	7	5	4	6	7
24	4	5	4	4	3	2
25	5	7	6	8	4	5
26	2	9	4	5	2	7
27	4	1	6	5	4	4
28	6	3	4	6	4	8
29	8	5	7	7	6	8
30	7	4	7	5	4	6
31	6	4	5	7	7	3
32	1	4	1	2	5	3
33	7	5	5	7	9	7
34	4	8	7	6	7	9
35	3	4	5	4	2	6
36	5	4	3	2	2	4
37	4	3	6	3	4	2
38	1	3	2	1	0	5
39	3	6	5	5	3	4
40	6	5	6	3	3	6

Child	A	L	FA	FL	MA	ML	Child	A	L	FA	FL	MA	ML
41	6	1	4	2	4	5	71	7	8	5	7	7	5
42	4	4	2	1	3	2	72	4	6	6	8	3	3
43	5	8	3	3	5	5	73	7	3	9	4	7	7
44	4	4	5	8	7	7	74	7	4	7	7	10	8
45	6	1	9	4	8	3	75	3	7	4	5	6	6
46	3	3	4	5	4	3	76	5	0	8	4	6	4
47	5	4	4	7	5	3	77	6	7	5	6	2	3
48	5	6	5	4	4	4	78	4	4	4	3	2	6
49	4	6	5	3	2	2	79	6	6	5	5	3	6
50	4	4	7	9	7	6	80	7	7	8	7	6	5
51	3	3	8	4	4	3	81	4	2	5	3	5	1
52	7	5	5	6	6	5	82	6	6	6	4	6	4
53	6	4	9	4	7	8	83	0	4	3	6	5	6
54	4	6	6	5	4	6	84	3	4	3	4	6	4
55	4	3	6	3	6	4	85	3	4	3	4	3	3
56	5	5	6	5	7	3	86	5	6	6	6	3	4
57	4	4	6	3	3	4	87	6	3	4	1	1	5
58	7	8	6	6	3	4	88	4	3	6	5	6	5
59	4	8	3	5	7	7	89	2	5	8	7	8	9
60	4	2	2	3	4	1	90	7	8	9	5	6	7
61	9	4	9	6	7	6	91	3	6	5	8	5	6
62	5	6	4	7	7	5	92	5	7	7	3	3	6
63	4	6	5	2	3	4	93	3	4	4	8	6	5
64	6	3	8	2	7	5	94	6	7	8	9	7	6
65	5	3	3	5	4	6	95	6	5	6	6	7	6
66	5	4	4	2	3	6	96	3	5	4	7	3	5
67	5	5	6	5	5	8	97	4	3	4	6	7	3
68	8	6	6	4	4	3	98	4	4	4	4	5	7
69	2	7	5	4	6	7	99	4	4	5	2	6	6
70	3	2	2	2	4	2	100	6	5	5	4	5	4

Part II

Bayesian Evaluation of Informative Hypotheses

This part of the book contains Chapters 3 through 6. In Chapter 3 the foundation of Bayesian evaluation of informative hypotheses is elaborated via the introduction of the density of the data, the prior distribution, the posterior distribution, the Bayes factor, and posterior model probabilities. The comparison of two independent means constitutes a very simple context within which all concepts are explained. Furthermore, there will be attention for sensitivity analysis, a first evaluation of the approach proposed, and an illustration using the software package BIEMS. Chapter 3 is very accessible because all technical material has been placed in appendices that are included to satisfy the curiosity of readers with above average interest and skill in the area of statistics.

Chapter 4 presents a full and accessible account of the evaluation of informative hypotheses in the context of the J group ANOVA model. The ease with which the approach proposed can be applied by behavioral and social scientists is illustrated using applications of the software package BIEMS.

In Chapter 5, error probabilities and a simple procedure that can be used for sample size determination are introduced. Subsequently, both are used for the evaluation of the performance of Bayesian evaluation of informative hypotheses in the context of the univariate normal linear model using the examples introduced in Chapter 1.

Chapter 6 contains a further discussion of error probabilities and presents a more elaborate procedure for sample size determination. Subsequently, both are used for the evaluation of the performance of the Bayesian approach in the context of the multivariate normal linear model using the examples introduced in Chapter 2.

42

Symbol Description

Chapter 3

μ_j — The mean in group j.

σ^2 — The residual variance.

y_i — The score of person i on dependent variable y.

$\overline{y}_j$ — The sample average of y in group j.

H_0 — The null hypothesis.

H_a — The alternative hypothesis.

H_m — An informative hypothesis.

i — Index for person 1 to N.

j — Index for group 1 to 2.

m — Index for hypothesis 1 to M.

μ_0 — Mean of the prior distribution of μ_j.

τ_0^2 — Variance of the prior distribution of μ_j.

ν_0 — Degrees of freedom of the prior distribution of σ^2.

σ_0^2 — Scale parameter of the prior distribution of σ^2.

$h(\mu_j|H_a)$ — Unconstrained prior distribution of μ_j.

$h(\sigma^2)$ — Prior distribution of σ^2.

$\mu_{j,N}$ — Mean of the posterior distribution of μ_j.

$\tau_{j,N}^2$ — Variance of the posterior distribution of μ_j.

c_m — The complexity of H_m.

f_m — The fit of H_m.

BF_{ma} — The Bayes factor of H_m versus H_a.

PMP_m — The posterior model probability of H_m.

Chapter 4

d — Effect size used in the specification of an informative hypothesis.

e — Effect size used in the specification of a population.

R_m — Matrix used to create inequality constraints.

r_m — Vector used to include effect sizes in inequality constraints.

S_m — Matrix used to create equality constraints.

s_m — Vector used to include effect sizes in equality constraints.

Chapter 5

$Z(y_i)$ — Standardized score of person i on the dependent variable.

$Z(x_{ki})$ — Standardized score of person i on the k-th predictor.

β_k — Regression coefficient relating the kth predictor x to the dependent variable y.

p-value — The probability of exceedance.

R^2 — Proportion of explained variance.

Chapter 6

y_{pi} — The score of person i on the p-th dependent variable y.

α_m — Probability of incorrectly rejecting H_m.

L — Lower bound of the interval of indecision.

U — Upper bound of the interval of indecision.

ρ — Correlation between (the residuals of) two dependent variables.

μ_{pj} — The mean in group j with respect to the p-th dependent variable y.

β_{pk} — Regression coefficient relating the k-th predictor x to the p-th dependent variable y.

Chapter 3

An Introduction to Bayesian Evaluation of Informative Hypotheses

3.1 Introduction

In this chapter Bayesian evaluation of informative hypotheses is introduced. The main concepts involved are explained using a simple two group ANOVA model and the hypotheses: $H_0 : \mu_1 = \mu_2$, $H_1 : \mu_1 > \mu_2$, $H_2 : |\mu_1 - \mu_2| < 1$, and $H_a : \mu_1, \mu_2$. The explanation is based on figures and descriptions to increase the accessibility of this chapter for behavioral and social scientists. The more statistically oriented reader is referred to Appendices 3A and 3B

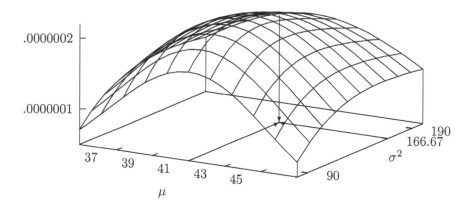

FIGURE 3.1: Density of the data.

and Chapter 10 in which the statistical foundations of Bayesian evaluation of informative hypotheses are presented.

This chapter introduces: the density of the data, prior and posterior distributions, complexity, fit, Bayes factors and posterior model probabilities, sensitivity to the prior distribution, and performance of the Bayes factor. The chapter concludes with a discussion that also provides further reading in the area of Bayesian statistics, that is, references discussing Bayesian statistics in general.

Appendix 3C illustrates how the software package BIEMS (Mulder, Hoijtink, and de Leeuw, In Press) can be used for the evaluation of informative hypotheses. Such illustrations will return in Chapters 4, 5, and 6 for BIEMS, and Chapter 7 and 8 for other packages that can be used for the evaluation of informative hypotheses. As will be shown, only a few commands are needed in order to be able to evaluate informative hypotheses.

3.2   Density of the Data, Prior and Posterior

3.2.1   Density of the Data

The density of the data is an important concept both in classical and Bayesian statistics. Imagine a population consisting of one group of persons whose ages have an unknown average μ and an unknown variance σ^2. Suppose a sample of four ages was obtained from this population: 26, 36, 46, and 56. The density of the data, $y = [26, 36, 46, 56]$ given μ and σ^2, will be denoted by $f(y \mid \mu, \sigma^2)$. The density of the data is a number that can be computed for every value of μ and σ. Given a sample mean of 41 and variance of 166.67, it is rather likely that the data come from a population with $\mu = 41$ and $\sigma^2 = 166.67$. This is reflected by the density of the data because it is at its maximum if μ and σ^2 are equal

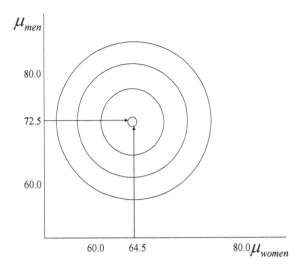

FIGURE 3.2: Two-dimensional representation of the density of the data.

to the sample mean and variance, respectively. This can be verified in Figure 3.1, where the density of the data is plotted for varying values of μ and σ^2. The three arrows indicate the point where $f(\boldsymbol{y} \mid \mu, \sigma^2)$ is at its maximum. As can be seen, the larger the deviation of μ from the sample mean and the larger the deviation of σ^2 from the sample variance, the smaller the density of the data.

The density of the data can be interpreted as the support in the data for specific values of μ and σ^2. In Figure 3.1 the support is largest for $\mu = 41$ and $\sigma^2 = 166.67$ and smaller for other combinations of μ and σ^2. The density of the data can not only be computed for a population consisting of one group, but also for a population containing multiple groups. Consider, for example, a population consisting of men and women. The variable of interest is weight. A sample obtained from this population consisted of four men with weights of 65, 70, 75, and 80 kilos (an average of 72.5 with a variance of 41.60), and four women with weights of 57, 62, 67, and 72 kilos (an average of 64.5 with a variance of 41.60). Figure 3.2 contains a two-dimensional representation of the density of the data $f(\boldsymbol{y} \mid \mu_{men}, \mu_{women}, \sigma^2)$ for varying values of μ_{men} and μ_{women}. The population variance σ^2 is ignored for the moment (it does not fit into a two dimensional figure already containing μ_{men} and μ_{women}) and assumed to have a known value of 41.60. The circle in the center represents the maximum of the density (the counterpart of the maximum in Figure 3.1), which is located at the sample mean of μ_{men} and μ_{women}. The circles around the center circle are called iso-density contours. Combinations of values of μ_{men} and μ_{women} on the same circle have the same density. The further a circle is located from the center circle, the smaller the density. In the sequel, densities will be visually presented by means of one circle. This will be the 95% iso-density contour, that is, the circle within which 95% of the density at hand is located. Section 3.6.1 in Appendix 3A contains the details of the density of the data for the J group ANOVA model.

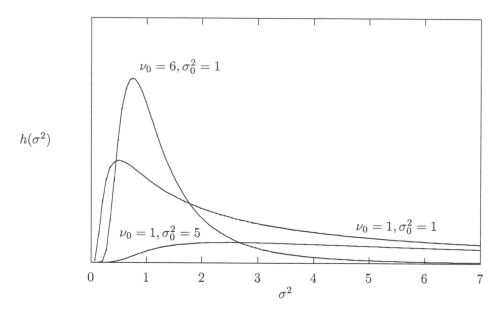

$h(\sigma^2)$

$\nu_0 = 6, \sigma_0^2 = 1$

$\nu_0 = 1, \sigma_0^2 = 5$

$\nu_0 = 1, \sigma_0^2 = 1$

σ^2

FIGURE 3.3: The prior distribution for σ^2.

3.2.2   Unconstrained Prior Distribution

The prior distribution is an important part of Bayesian statistical inference. The prior distribution represents the knowledge with respect to the parameters of a statistical model that is available *before* the data are observed, that is, what is *a priori* the support for each combination of parameter values. Note the similarity with the definition of the density of the data: the support *in the data* for each combination of parameter values.

In a population consisting of two groups of persons with a certain weight as described in the previous section, the parameters of the statistical model are μ_1, μ_2, and σ^2. The prior distribution for these parameters will be denoted by $h(\boldsymbol{\mu} \mid H_a)h(\sigma^2)$. Note that $\boldsymbol{\mu}$ is a vector containing both μ_1 and μ_2. Note furthermore that $h(\boldsymbol{\mu} \mid H_a)$ is the prior support for all possible combinations of μ_1 and μ_2 under the hypothesis $H_a : \mu_1, \mu_2$, that is, the means are unconstrained. Note finally that $h(\sigma^2)$ holds for unconstrained and constrained hypotheses. This is stressed by the absence of a conditioning on the unconstrained hypothesis H_a or an informative hypothesis H_m.

3.2.2.1   The Prior for the Within Group Variance

The prior for σ^2 is visualized in Figure 3.3. It is a so-called scaled inverse chi-square distribution (see Gelman, Carlin, Stern, and Rubin (2004, pp. 574, 580), or Appendix 10C). The prior support for each value of σ^2 is indicated by the height of the curve. The shape of this distribution is determined by the degrees of freedom ν_0 which determine the vagueness of the curve and the scale parameter σ_0^2 which determines its location. The choice of values for ν_0 and σ_0^2 are discussed in Section 3.4. If the goal is to select the best of a set of informative hypotheses, the only requirement is that the prior distribution of σ^2 is vague.

Property 3.1: If the prior distribution of σ^2 is vague, it hardly influences the evaluation of informative hypotheses.

Area in the lower right-hand triangle is in agreement with $H_1 : \mu_1 > \mu_2$

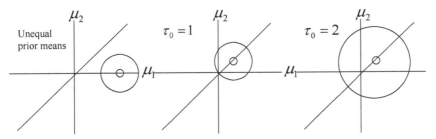

Area within the diagonal band is in agreement with $H_2 : |\mu_1 - \mu_2| < 1$

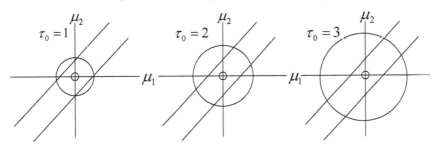

FIGURE 3.4: The prior distribution for μ_1 and μ_2.

3.2.2.2   The Prior for the Means

When the goal is to evaluate informative hypotheses, the specification of the prior distribution for μ_1 and μ_2 is of crucial importance. Without constraints, that is, for $H_a : \mu_1, \mu_2$, the prior is

$$h(\boldsymbol{\mu} \mid H_a) = h(\mu_1, \mu_2 | H_a) \sim \mathcal{N}\left(\begin{bmatrix} \mu_0 \\ \mu_0 \end{bmatrix}, \begin{bmatrix} \tau_0^2 & \tau_{00} \\ \tau_{00} & \tau_0^2 \end{bmatrix}\right) = \mathcal{N}(\boldsymbol{\mu}_0, \boldsymbol{T}_0), \qquad (3.1)$$

which is a bivariate normal distribution with means μ_0, variances τ_0, and covariance τ_{00}. A crucial feature of the prior is that both μ_1 and μ_2 have the same prior mean μ_0 and variance τ_0^2. The choice for this feature is motivated below and elaborated in Chapters 4, 7, and 10. In the remainder of this chapter $\tau_{00} = 0$ which implies that a priori the means are independent:

$$h(\mu_1, \mu_2 \mid H_a) \sim h(\mu_1 | H_a) h(\mu_2 | H_a), \qquad (3.2)$$

where $h(\mu_j | H_a) = \mathcal{N}(\mu_0, \tau_0^2)$ for $j = 1, 2$.

    Prior distributions constructed using different values of μ_0 and τ_0^2 can be found in Figure 3.4. As for σ^2, the prior support for each value of μ is determined by its prior density. The iso-density contours (the larger circles) in Figure 3.4 contain 95% of the prior distribution of μ_1 and μ_2 (the larger τ_0^2 the wider the circle), the maximum of the prior distribution is indicated by the smaller circle. The diagonal line represents $\mu_1 = \mu_2$. The diagonal band represents $|\mu_1 - \mu_2| < 1$.

    Suppose the informative hypothesis under investigation is $H_1 : \mu_1 > \mu_2$. In this case the prior specified for H_a is neutral with respect to H_1 and its complement $H_{1_c} : \mu_1 < \mu_2$ because 50% of the resulting prior is in agreement with each of these hypotheses. As can be verified in Figure 3.4, this is true for any value of μ_0 and τ_0^2 (top row, middle and right-hand figures). This is an important property: if only inequality constraints are used to specify the informative hypotheses, the prior distribution (3.2) is neutral with respect to the informative hypothesis under investigation. Furthermore, this property is independent of the value of μ_0 and τ_0^2. Note that if μ_1 and μ_2 have different prior means (see the left-hand figure in the top row), the prior distribution is not neutral because it mainly supports H_1 and has little support for H_{1_c}. A prior distribution that is neutral can be used to quantify the complexity of an inequality constrained hypothesis under investigation. As can be seen, H_1 and H_{1_c} are equally complex and, loosely spoken, each is half as complex as H_a. This is adequately represented by the fact that 50% of the prior distribution is in agreement with H_1 and H_{1_c}, and 100% is in agreement with H_a. This leads to Property 3.2:

Property 3.2: If only inequality constraints are used to specify informative hypotheses, the prior distribution for H_a can be specified such that it is neutral with respect to the informative hypotheses under investigation, that is, it adequately quantifies the complexity of the informative hypotheses under investigation. For these informative hypotheses, the quantification of complexity is independent of the choice of the prior mean μ_0 and variance τ_0^2.

All publications with respect to the evaluation of informative hypotheses in the context of ANOVA models have implicitly or explicitly used Property 3.2. The interested reader is referred to Klugkist, Kato, and Hoijtink (2005); Klugkist, Laudy, and Hoijtink (2005); Klugkist and Hoijtink (2007); Van Deun, Hoijtink, Thorrez, van Lommel, Schuit, and van Mechelen (2009); and Kuiper and Hoijtink (2010). The interested reader is also referred to Mulder (Unpublished) for a recent critical appraisal.

Suppose the only informative hypothesis under investigation is $|\mu_1 - \mu_2| < 1$, that is, μ_1 and μ_2 are about equal to each other. This states that μ_1 and μ_2 are within a distance of 1 from each other. This is a hypothesis specified using an effect size as was discussed in Section 1.2.3. Figure 3.4 illustrates the effect of the size of τ_0^2 on the support in the prior for $H_2 : |\mu_1 - \mu_2| < 1$. As can be seen in the bottom row, the larger τ_0^2, the smaller the proportion of the prior distribution in agreement with H_2. The implication is that the choice of τ_0^2 has an effect on the evaluation of informative hypotheses specified using equality or about equality constraints. The specification of a proper value for τ_0^2 is elaborated in Section 3.4. These considerations are summarized in Property 3.3:

Property 3.3:. If informative hypotheses are specified using equality or about equality constraints, the choice of τ_0^2 determines the support in the prior for these hypotheses and thus their complexity.

Property 3.3 implies that the concept of neutrality cannot be applied to hypotheses specified using equality or about equality constraints. However, an intuitively appealing property is obtained if (3.1) or (3.2) is used:

Property 3.4:. For informative hypotheses specified using equality or about equality constraints, the prior contains the same information for each of the means constrained to be equal.

3.2.3   Constrained Prior Distribution

The prior distribution for constrained hypotheses can be derived from the prior distribution of the corresponding (that is, having the same number of means) unconstrained hypothesis. Section 3.6.2 in Appendix 3A contains the details of this derivation for the J group ANOVA model. First of all, $h(\sigma^2)$ does not depend on whether the hypothesis at hand is constrained or not, that is, the prior distribution of the within group variance is the same for unconstrained and constrained hypotheses. The prior distribution of μ_1 and μ_2 for constrained hypotheses can be derived from the prior distribution of the corresponding unconstrained hypothesis. As can be seen in the center figure of the top row in Figure 3.4, the prior distribution of $H_1 : \mu_1 > \mu_2$ is just the part of the prior distribution of $H_a : \mu_1, \mu_2$ below the diagonal line. It is important to note that 50% of H_a is in agreement with H_1. Similarly, as can be seen in the left-hand figure of the bottom row in Figure 3.4, the prior distribution of $H_2 : |\mu_1 - \mu_2| < 1$ is just the part of the prior distribution of $H_a : \mu_1, \mu_2$ within the diagonal band. Here, about 90% of the prior distribution of H_a is in agreement with H_2. In the right-hand figure on the bottom row, this is about 30%.

This manner of prior construction is in line with the principle of Leucari and Consonni (2003) and Roverate and Consonni (2004): "If nothing was elicited to indicate that the two priors should be different, then it is sensible to specify [the prior of constrained hypotheses] to be, ..., as close as possible to [the prior of the unconstrained hypothesis]. In this way the resulting Bayes factor should be least influenced by dissimilarities between the two priors due to differences in the construction processes, and could thus more faithfully represent the strength of the support that the data lend to each [hypothesis]". Note that in this book the Bayes factor will be used to quantify the relative support in the data for pairs of hypotheses. The Bayes factor will be introduced in Section 3.3.1.

The proportion of the unconstrained prior distribution in agreement with the constrained hypothesis will be called "the complexity of a constrained hypothesis," to be denoted by c_m. The smaller c_m, the more precise H_m, that is, the smaller the admissible space of parameter values. This is an important property that will reappear in Chapters 4, 7, and 10.

Property 3.5: The complexity of a constrained hypothesis will be denoted by c_m, which is the proportion of the unconstrained prior distribution in agreement with H_m.

In the right-hand figure in the top row of Figure 3.4, $c_1 = .50$, in the left-hand figure in the bottom row $c_2 \approx .90$, in the bottom right-hand figure $c_2 \approx .30$. Property 3.5 breaks down if a hypothesis is specified using exact equalities, for example, $H_0 : \mu_1 = \mu_2$, because the proportion of the unconstrained prior distribution in agreement with H_0 is 0. As will be elaborated in Chapter 10, this situation can be handled if H_0 is reformulated as $H_2 : |\mu_1 - \mu_2| < \eta$, where η is so small that it is no longer possible to distinguish between H_0 and H_2. Note that H_2 can be visualized using a really small diagonal band in the bottom row of Figure 3.4.

3.2.4   Unconstrained Posterior Distribution

The posterior distribution combines the information with respect to μ_1, μ_2, and σ^2 in the density of the data and the prior distribution (see Section 3.6.3 in Appendix 3A for a statistical elaboration). Stated otherwise, the posterior distribution quantifies the support for specific values of μ_1, μ_2, and σ^2 by combining the support in the density of the data and the prior distribution. To facilitate the introduction of the posterior distribution, in the remainder of this section it is assumed that the value of σ^2 is known.

As can be seen in each of the four figures in Figure 3.5, the prior distribution of μ_1 and μ_2 has a mean on the line $\mu_1 = \mu_2$, that is, μ_1 and μ_2 have the same prior mean. Note that 95% of the prior distribution is contained within the largest circle, that is, the prior variance τ_0^2 of μ_1 and μ_2 is relatively large. The smaller circles represent the density of the data. The mean is given by the sample means $\bar{y}_1$ and $\bar{y}_2$ and the variance by σ^2/N_1 and σ^2/N_2. The dashed circles represent the posterior distribution of μ_1 and μ_2. The posterior

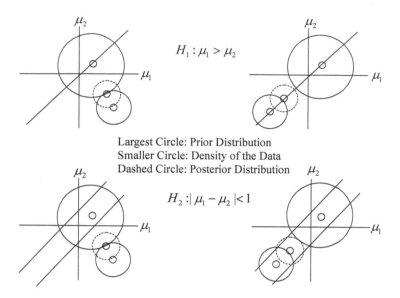

$$H_1 : \mu_1 > \mu_2$$

Largest Circle: Prior Distribution
Smaller Circle: Density of the Data
Dashed Circle: Posterior Distribution

$$H_2 :| \mu_1 - \mu_2 |< 1$$

FIGURE 3.5: The posterior distribution for μ_1 and μ_2.

distribution combines the information in density and prior. The posterior mean of μ_j for $j = 1, 2$ is

$$\mu_{j,N} = \frac{\frac{\mu_0}{\tau_0^2} + \frac{\bar{y}_j}{\sigma^2/N_j}}{\frac{1}{\tau_0^2} + \frac{1}{\sigma^2/N_j}}, \tag{3.3}$$

that is, the posterior mean is the weighted average of the prior mean and the mean of the density of the data, where the weights are the variance of the prior and the variance of the density of the data, respectively (Gelman, Carlin, Stern, and Rubin, 2004, pp. 48–49). In Figure 3.5 it can be seen that the posterior mean is located between the mean of the density and the prior mean. It can furthermore be observed that it is located closer to the mean of the density because the variance of the density is smaller than the variance of the prior. The posterior variance of μ_j for $j = 1, 2$ is (Gelman, Carlin, Stern, and Rubin, 2004, pp. 48–49)

$$\tau_{j,N} = \frac{1}{\frac{1}{\tau_0^2} + \frac{1}{\sigma^2/N_j}}. \tag{3.4}$$

As can be seen in Figure 3.5, the posterior variance is smaller than the variance of the density and the prior. This is because the combined information with respect to μ_1 and μ_2 is larger than the information in each of the two underlying components. The following property can be derived from (3.3) and (3.4):

Property 3.6: For $\tau_0^2 \to \infty$, both the $\mu_{j,N}$ and the $\tau_{j,N}$ are independent of the prior distribution, that is, the posterior distribution is completely determined by the data.

The top row of Figure 3.5 addresses $H_1 : \mu_1 > \mu_2$. The left-hand figure displays a situation where the data are in agreement with H_1, that is, the sample means are in agreement with H_1. As can be seen, about 99% of the posterior distribution (the dashed circle contains 95% of the posterior distribution) is located in the parameter space admitted by H_1. The right-hand figure in the top row displays the situation where $\overline{y}_1 = \overline{y}_2$. As can be seen, for this situation, 50% of the posterior distribution is located in the parameter space admitted by H_1. The bottom row of Figure 3.5 addresses $H_2 : |\mu_1 - \mu_2| < 1$. As can be seen in the left-hand figure, about 1% of the posterior distribution is in agreement with H_2. In the right-hand figure, about 95% is in agreement with H_2. This leads to another property that will be used throughout this book:

Property 3.7: The fit of a constrained hypothesis, to be denoted by f_m, is the proportion of the unconstrained posterior distribution in agreement with H_m.

Properties 3.6 and 3.7 are at the basis of the work presented in Klugkist, Kato, and Hoijtink (2005); Klugkist, Laudy, and Hoijtink (2005); Klugkist and Hoijtink (2007); Van Deun, Hoijtink, Thorrez, van Lommel, Schuit, and van Mechelen (2009); and Kuiper and Hoijtink (2010). Like Property 3.5, Property 3.7 breaks down if the informative hypothesis is specified using exact equalities, for example, $H_0 : \mu_1 = \mu_2$. Again, as is elaborated in Chapter 10, this situation can be handled if H_0 is reformulated as $H_2 : |\mu_1 - \mu_2| < \eta$, where η is so small that it is no longer possible to distinguish between H_0 and H_2.

3.3   Bayesian Evaluation of Informative Hypotheses

3.3.1   The Bayes Factor

The Bayes factor is a Bayesian model selection criterion. Model selection can be applied if there are two or more hypotheses under investigation and the goal is to determine the support in the data for each of these hypotheses. The Bayes factor of a constrained hypothesis H_m versus the unconstrained hypothesis H_a is defined as

$$BF_{ma} = f_m/c_m. \tag{3.5}$$

The derivation of this equation can be found in Section 3.7.1 in Appendix 3B and, for example, Klugkist, Laudy, and Hoijtink (2005). As can be seen, the Bayes factor accounts for both the fit and the complexity of a hypothesis. The larger the fit and the smaller the complexity, the higher the Bayes factor. In this sense the Bayes factor of a constrained hypothesis versus the unconstrained hypothesis is consistent with the argument that Sober (2002) makes in favor of plausibility, that is, support of the data combined with parsimony.

If $BF_{ma} = 5$, this implies that after observing the data, the support for H_m is five times as large as the support for H_a. This is consistent with the argument that Carnap (1952) makes in favor of degree of confirmation of a scientific hypothesis and with the notion of evidence in favor of a hypotheses, which is the central concept in Royall (1997). Stated otherwise, the Bayes factor is a quantification of the degree of evidence in the data in favor of each of a pair of hypotheses. As is more elaborately discussed in Chapters 5 and 6, there are no general guidelines for the interpretation of Bayes factors analogous to the benchmark of .05 that is often used to interpret a p-value. However, note that most researchers will agree that $BF_{ma} = 10$, that is, the support for H_m is ten times larger than the support for H_a, is convincing evidence in favor of H_m. Similarly, $BF_{ma} = 1.4$ is neither convincing evidence in favor of H_m nor convincing evidence against H_a. Note that 1 is an important reference value for the interpretation of the Bayes factor. If $BF_{ma} > 1$, the support in the data is, however slightly, in favor of H_m. If $BF_{ma} < 1$, the support in the data is, however slightly, in favor of H_a.

Looking at the left-hand side of Figure 3.5, it can be seen that the density of the data does support the prior of H_1 and does not support the prior of H_2. This translates into $BF_{1a} \approx .99/.50 \approx 2$ and $BF_{2a} \approx .01/.40 \approx .025$, where, .99 and .01 are the proportions of the unconstrained posterior in agreement with H_1 and H_2, respectively, and, .50 and .40 the corresponding prior proportions. Stated otherwise after observing the data, H_1 is twice as likely as H_a and H_2 is .025 as likely as H_a. The Bayes factor of two constrained hypotheses H_m and $H_{m'}$ can be written as

$$BF_{mm'} = BF_{ma}/BF_{m'a} = \frac{f_m}{c_m} \Big/ \frac{f_{m'}}{c_{m'}}. \tag{3.6}$$

For the example at hand, this leads to $BF_{12} = BF_{1a}/BF_{2a} = 2/.025 = 80$, that is, H_1 is 80 times as likely as H_2 after observing the data. Note that $BF_{mm'}$ provides a relative degree of evidence of the kind "which is the best of the hypotheses under investigation" and not an absolute degree of evidence of the kind "how good is this hypothesis." However, as was also elaborated in Section 2.6, as long as a constrained hypothesis receives more support from the data than the unconstrained hypothesis, the problem that "the best of a set of poorly fitting hypotheses is selected" can be avoided.

Looking at the right-hand side of Figure 3.5, it can be seen that $BF_{1a} = .50/.50 = 1$ and $BF_{2a} \approx .95/.40 \approx 2.375$, leading to a $BF_{12} = 1/2.375 \approx .42$. Stated otherwise, after observing the data, H_2 is about 2.375 times as likely as H_1.

The Bayes factor BF_{m,m_c} of H_m versus its complement H_{m_c} will be presented in Section 3.4.4. The interested reader is referred to Lavine and Schervish (1999) for a more elaborate exposition of the interpretation of the Bayes factor: "Bayes factors: What they are and what they are not." A standard reference to Bayes factors is the review paper by Kass and Raftery (1995). Van de Schoot, Mulder, Hoijtink, van Aken, Dubas, Orobio de Castro, Meeus, and Romeijn (In Press) discuss the evaluation of Bayes factors computed for informative hypotheses. In the next chapter, examples of the application of the Bayes factor when the goal is to evaluate informative hypotheses in the context of ANOVA are extensively discussed. Using the examples introduced in Chapter 1, this will go beyond the two group context that is used here to introduce Bayesian model selection.

3.3.2   Posterior Model Probabilities

It takes more than a glance to evaluate two or more Bayes factors. Consider again the first example from the previous section: $BF_{1a} = 2$, $BF_{2a} = .025$, and $BF_{12} = 80$. The correct conclusion is that the support in the data is largest for H_1, followed by H_a and smallest for H_2. To be more precise, H_1 is 2 times as likely as H_a, which in turn is 40 times as likely as H_2.

Posterior model probabilities (PMPs) are an alternative way to present the information in a set of Bayes factors. In this book the point of departure will always be the computation of Bayes factors with respect to the unconstrained hypothesis, that is, $BF_{1a}, \ldots, BF_{Ma}$. These can be translated into PMPs for all the hypotheses under investigation, including H_a, using

$$PMP_m = \frac{BF_{ma}}{1 + \sum_m BF_{ma}} \text{ for } m = 1, \ldots, M \tag{3.7}$$

and

$$PMP_a = \frac{1}{1 + \sum_m BF_{ma}}. \tag{3.8}$$

For the example at hand, $PMP_1 = 2/(2+.025+1) = .66$, $PMP_2 = .025/(2+.025+1) = .01$ and $PMP_a = 1/(2+.025+1) = .33$. These numbers reveal at a glance the support in the data for the hypotheses under investigation. Note that in (3.7), (3.8), and (3.9) $PMPs$ are computed assuming that a priori each hypothesis under investigation is equally likely. This makes the $PMPs$ straightforward translations of a set of Bayes factors to numbers on a scale running from 0 to 1 that express the support in the data for each hypothesis under investigation. If H_a is not included in the set of hypotheses under investigation, $PMPs$ are computed via

$$PMP_m = \frac{BF_{ma}}{\sum_m BF_{ma}} \text{ for } m = 1, \dots, M. \qquad (3.9)$$

Using this formula we obtain for the example at hand that $PMP_1 = 2/(2 + .025) = .99$ and $PMP_2 = .025/(2 + .025) = .01$.

To avoid misinterpretation of the $PMPs$, the following property must be highlighted:

Property 3.8:. A PMP is a quantification of the support in the data for the hypothesis at hand on an interval that runs from 0 to 1, in which fit and complexity of the hypothesis are accounted for. The PMP is a logical probability (Carnap, 1950; Jeffreys, 1961; Jaynes, 2003). The PMP is *not* a classical frequentist probability which, like the probability that a coin falls on its tail, or a dice showing the number four after it has been thrown. Consequently, it cannot be interpreted a such. This is further highlighted by a quote from Jeffreys (1961) used as the introduction to Chapter 9 from Jaynes (2003): "The essence of the present theory is that no probability, direct, prior, or posterior, is simply a frequency."

3.4 Specifying the Parameters of Prior Distributions

The unconstrained prior distribution used in this chapter for the two group ANOVA model is

$$h(\mu_1, \mu_2 \mid H_a)h(\sigma^2) \sim h(\mu_1 \mid H_a)h(\mu_2 \mid H_a)h(\sigma^2), \qquad (3.10)$$

where $h(\mu_j \mid H_a) = \mathcal{N}(\mu_0, \tau_0^2)$ for $j = 1, 2$ and $h(\sigma^2) = \text{Inv-}\chi^2(\nu_0, \sigma_0^2)$. Before Bayes factors and $PMPs$ can be computed, values must be specified for μ_0, τ_0^2 (the parameters of the normal prior distribution), and ν_0 and σ_0^2 (the parameters of the scaled inverse chi-square distribution), that is, values must be computed for the prior mean and variance, and, for the prior degrees of freedom and scale parameter.

3.4.1 Hypotheses Specified Using Inequality Constraints

As was formulated in Property 3.1, the evaluation of informative hypotheses is unaffected by the choice of $h(\sigma^2)$ as long as it is vague (see Section 3.7.2.1 in Appendix 3B for further elaboration). Using this result, a default choice is $\nu_0 = 0$ and an arbitrary value for σ_0^2, which renders the Jeffreys prior for a variance $h(\sigma^2) \propto 1/\sigma^2$ (see Appendix 10C). Default means that human intervention or choices are not required to determine the prior parameters. Objective means that Bayes factors based on this prior distribution are essentially independent of this prior distribution.

As was formulated in Property 3.2, as long as informative hypotheses are specified using only inequality constraints c_i is independent of the choice of μ_0 and τ_0^2 (see Section 3.7.2.2 in Appendix 3B for further elaboration). As was formulated in Property 3.6, for $\tau_0^2 \to \infty$, f_i is independent of $h(\mu_1, \mu_2 \mid H_a)$. Stated otherwise, for inequality constrained hypotheses, default and objective Bayes factors can be obtained using $\mu_0 = 0$, $\tau_0^2 \to \infty$ (Klugkist, Kato, and Hoijtink, 2005; Hoijtink, Unpublished).

3.4.2   Hypotheses Specified Using (About) Equality Constraints

As was formulated in Property 3.3, τ_0^2 determines the support in the prior for hypotheses specified using (about) equality constraints. For $\tau_0^2 \to \infty$, the Bayes factor BF_{0a} also goes to infinity, that is, the larger the prior variance, the larger the support for H_0 (see Section 3.7.2.2 in Appendix 3B for further elaboration). Stated otherwise, if one of the hypotheses under investigation is specified using equality or about equality constraints, the specification of ν_0 and σ_0^2 can, due to Property 3.1, be analogous to the the specification given in the previous section, but due to Property 3.3 there are no longer simple choices for μ_0 and τ_0^2.

A large part of the statistical research in the area of Bayesian evaluation of informative hypotheses has focused on the specification of values for μ_0 and τ_0^2 (see, for example, Klugkist, Laudy, and Hoijtink, 2005; Kuiper and Hoijtink, 2010; Mulder, Hoijtink, and Klugkist, 2010; and Mulder, Hoijtink, and de Leeuw, In Press). To avoid very technical elaborations, here only the main feature of the resulting specification is presented (see Sections 3.6.2.1 and 3.6.2.2 in Appendix 3A for an elaboration of the two specification methods that are currently used in Bayesian software for the evaluation of informative hypotheses). The specification of μ_0 and τ_0^2 is based on the idea that μ_0 should closely relate to the overall mean (that is, ignoring group membership) and that τ_0^2 should represent the uncertainty with respect to the overall mean in a very small subset of the data. A straightforward translation of this idea to a two group ANOVA would be to use $\mu_0 = \overline{y}$, that is, specify that the prior mean is equal to the sample average of y. Furthermore, one could use $\tau_0^2 = 1/2s^2$, that is, the prior variance is equal to the squared standard error of the mean computed as if the sample size was 2 instead of N. Note that s^2 denotes the sample variance, that is, $s^2 = 1/N \sum_i (y_i - \overline{y})^2$. A first impression of the performance of Bayes factors based on this principle is given in Section 3.4.4. Further elaborations of the performance follow in Chapters 4, 5, and 6.

3.4.3   Estimation of Complexity and Fit

Once the prior distribution has been specified, it can be used to estimate c_m for each informative hypothesis. This is achieved using a sample from $h(\boldsymbol{\mu} \mid H_a)$. The proportion of the sample in agreement with H_m is an estimate of c_m. As can be seen in Figure 3.5, a large sample of μ_1 and μ_2 from the unconstrained prior distribution in the top left-hand figure would render $\mu_1 > \mu_2$ in 50% of the cases, and $\mu_1 < \mu_2$ in 50% of the cases. Stated otherwise, the resulting estimate of c_1 is .50. A large sample from the unconstrained prior distribution in the bottom left-hand figure would render an estimate of about .40 for c_2. In a similar vein, the posterior distribution can be used to estimate f_m for each informative hypothesis.

Sampling from both the prior and posterior distribution can be achieved using Markov chain Monte Carlo methods. The interested reader is referred to Klugkist and Mulder (2008) for an accessible account in the context of ANOVA and Chapter 10 for a general elaboration of these approaches.

3.4.4   Performance of the Bayes Factor

To get a first impression of the performance of the Bayes factor, a small study was executed. In Table 3.1, five three Bayes factors are presented:

1. BF_{1a}, which compares $H_1 : \mu_1 > \mu_2$ with $H_a : \mu_1, \mu_2$,

2. BF_{0a}, which compares $H_0 : \mu_1 = \mu_2$ with H_a,

3. $BF_{10} = BF_{1a}/BF_{0a}$, which compares H_1 with H_0.

TABLE 3.1: Performance of the Bayes Factor for the Comparison of Two Means

	$e=0$	$e=.2$	$e=.5$	$e=.8$	$e=1$
		N = 20 Per Group			
BF_{1a}	1.00	1.44	1.85	1.98	2.00
BF_{0a}	2.78	2.35	.98	.22	.06
BF_{10}	.36	.61	1.89	9.00	33.33
		N = 50 Per Group			
BF_{1a}	1.00	1.67	1.98	2.00	2.00
BF_{0a}	3.99	2.56	.25	.01	.00
BF_{10}	.25	.65	7.92	200	∞

The Bayes factors are computed with the software package BIEMS (see Section 9.2.2). Five data sets are constructed. In each data set the sample standard deviation is 1 for each group. The sample mean in group 2 is 0, and the sample mean in group 1 is 0, .2, .5, .8, and 1.0, respectively, for five different data sets. This renders sample values of Cohen's effect size measure e (see Chapter 1) equal to sample mean in group 1 because $e = (\bar{y}_1 - 0)/1 = \bar{y}_1$. Note that, in this book, d is used to denote an effect size used in the specification of an informative hypothesis, and e is used to denote an effect size in a population from which data are generated. The effect size measures e are given in the top row of Table 3.1. Finally, the top panel in Table 3.1 refers to a sample size of 20 per group, and the bottom panel to a sample size of 50 per group. How to obtain BF_{1a} and BF_{0a} for $d = .5$ and $N = 20$ using BIEMS is illustrated in Appendix 3C.

Let us first take a look at the comparison of H_1 with H_a. As can be seen on the line labeled BF_{1a} for $N = 20$ per group, the Bayes factor increases if the effect size increases. This is as it should be, because larger effect sizes provide more evidence in favor of H_1 than smaller effect sizes. The same can be observed on the corresponding line for $N = 50$ per group. The main difference is that there the Bayes factor increases faster. This too is as it should be because the larger the sample size, the smaller the probability that an effect is due to chance.

Two numbers deserve extra attention: the Bayes factor reported for an effect size of 0 and 1. The value 1 for $e = 0$ results because 50% of both the unconstrained prior and posterior are in agreement with H_1 resulting in a Bayes factor $BF_{1a} = .50/.50 = 1$. Stated otherwise, if prior and posterior support are equal, the Bayes factor has no preference for either hypothesis. The value 2 for $d = 1$ is the largest value that can be obtained for BF_{1a}. Note that the proportion of the unconstrained prior in agreement with H_1 is .50 and that this proportion does not depend on the data, that is, whether or not the data are in agreement with H_1, $c_1 = .50$. This number means that H_1 is half as complex as H_a. The value of f_1 does depend on the data: the more the data are in agreement with H_1, the better the fit, that is, the larger f_1 with a maximum value of 1.0. This implies that BF_{1a} is at most $1.0/.50 = 2$, which reflects that H_1 and H_a have the same fit, but that H_1 is half as complex as H_a, resulting in a Bayes factor of 2 in favor of H_1. An example of $BF_{1a} = 2$ can be found in Table 3.1 for $e = 1$ and $N = 20$.

Instead of BF_{1a} one could also compute the Bayes factor of a hypothesis with respect to its complement. This Bayes factor can be written as

$$BF_{mm_c} = \frac{f_m}{c_m} / \frac{1 - f_m}{1 - c_m}. \tag{3.11}$$

For the example at hand, $BF_{11_c} = (1/.50)/(0/.50) = \infty$, that is, although the support for H_1 is only twice as large as the support for H_a, the support for H_1 is infinitely larger than the support for H_{1_c}: not H_1.

As can be seen in the lines labeled BF_{0a}, the Bayes factor for $H_0 : \mu_1 = \mu_2$ decreases with increasing effect size, and decreases faster if the sample size is larger. This too is as it should be, because the larger the effect size, the larger the amount of evidence against H_0. The lines labeled BF_{10} contain the ratio of the two numbers directly above: BF_{1a} and BF_{0a}. As can be seen the larger the effect size, the larger the support in favor of $H_1 : \mu_1 > \mu_2$, and, for the larger sample size, the support is more pronounced. Again, this is in line with what would be required of a sensible criterion for the evaluation of informative hypotheses.

3.5 Discussion

In this chapter the main features of Bayesian evaluation of informative hypotheses have been introduced using a two group ANOVA model. A full account of the concepts and properties introduced in this chapter for the full multivariate normal linear model introduced in Chapter 2 are given in Chapter 10. The concepts and properties introduced are relevant for all models discussed in this book. The next chapter illustrates the evaluation of informative hypotheses for the J group ANOVA model. Chapters 5 and 6 contain illustrations for the univariate and multivariate normal linear model, respectively, and discuss the sample sizes that are needed in order to obtain reliable evaluations of informative hypotheses.

This book focuses on Bayesian evaluation of informative hypotheses. Readers interested in general introductions to Bayesian statistics are referred to Gelman, Carlin, Stern, and Rubin (2004), Gill (2002), Congdon (2003) and Lynch (2007). A one-chapter summary of the main features is provided by Hoijtink (2009). All the concepts introduced in this chapter: density of the data (see also Severini (2000)); prior (see also Kass and Wasserman, 1996) and posterior distribution; Bayes factors (see also Kass and Raftery (1995) and Lavine and Schervish (1999)); posterior model probabilities; and Markov chain Monte Carlo sampling, are discussed in these books.

3.6 Appendix 3A: Density of the Data, Prior and Posterior

3.6.1 Density of the Data

The density of the data for the J group ANOVA model is

$$f(\boldsymbol{y} \mid \boldsymbol{d}_1, \ldots, \boldsymbol{d}_J, \boldsymbol{\mu}, \sigma^2) =$$

$$\prod_i \frac{1}{\sqrt{2\pi\sigma^2}} \exp -\frac{1}{2} \frac{(y_i - \mu_1 d_{1i} - \ldots - \mu_J d_{Ji})^2}{\sigma^2}, \qquad (3.12)$$

where $\boldsymbol{y} = [y_1, \ldots, y_N]$, $\boldsymbol{\mu} = [\mu_1, \ldots, \mu_J]$, $\boldsymbol{d}_j = [d_{j1}, \ldots, d_{jN}]$, and $d_{ji} = 1$ if person i is a member of group j, and 0 otherwise. This density can be derived from the main equation of the ANOVA model

$$y_i = \mu_1 d_{1i} + \ldots + \mu_J d_{Ji} + \epsilon_i, \qquad (3.13)$$

combined with the assumption that $\epsilon_i \sim \mathcal{N}(0, \sigma^2)$.

3.6.2 Prior Distribution

For the ANOVA model specified in (3.13) the prior distribution for the parameters of a constrained hypothesis is obtained as

$$h(\boldsymbol{\mu}, \sigma^2 | H_m) = \frac{h(\boldsymbol{\mu}, \sigma^2 | H_a) I_{\boldsymbol{\mu} \in H_m}}{\int h(\boldsymbol{\mu}, \sigma^2 | H_a) I_{\boldsymbol{\mu} \in H_m} d\boldsymbol{\mu}, \sigma^2} = 1/c_m h(\boldsymbol{\mu}, \sigma^2 | H_a) I_{\boldsymbol{\mu} \in H_m}, \quad (3.14)$$

where c_m is the proportion of the unconstrained prior distribution in agreement with H_m, $I_{\boldsymbol{\mu} \in H_m}$ equals 1 if $\boldsymbol{\mu}$ is in agreement with the constraints of H_m and 0 otherwise,

$$h(\boldsymbol{\mu}, \sigma^2 | H_a) = h(\boldsymbol{\mu} | H_a) h(\sigma^2), \quad (3.15)$$

with

$$h(\boldsymbol{\mu} | H_a) = h\left(\begin{bmatrix} \mu_1 \\ \mu_2 \\ \cdots \\ \mu_J \end{bmatrix} \middle| II_a\right) \sim \mathcal{N}\left(\begin{bmatrix} \mu_0 \\ \mu_0 \\ \cdots \\ \mu_0 \end{bmatrix}, \begin{bmatrix} \tau_0^2 & \tau_{00} & \cdots & \tau_{00} \\ \tau_{00} & \tau_0^2 & \cdots & \tau_{00} \\ \cdots & \cdots & \cdots & \cdots \\ \tau_{00} & \tau_{00} & \cdots & \tau_0^2 \end{bmatrix}\right) = \mathcal{N}(\boldsymbol{\mu}_0, \boldsymbol{T}_0)$$

(3.16)

and

$$h(\sigma^2) \sim \text{Inv-}\chi^2(\nu_0, \sigma_0^2). \quad (3.17)$$

3.6.2.1 Prior Specification Method 1

For the J group ANOVA model, values for μ_0 and τ_0^2 can be specified using the method of Klugkist, Laudy, and Hoijtink (2005). They specify $\tau_{00} = 0$ and subsequently use the following three-step procedure to specify a prior for the μ_js:

- Compute the 99.7% confidence interval of μ_j for $j = 1, \ldots, J$, which is equal to $\bar{y}_j \pm 3 \times SE_j$. Note that $\bar{y}_j$ denotes the sample average of y in group j and SE_j the standard error of the mean in group j. The resulting intervals could, for example, be [−2,3] for group 1, [−1,5] for group 2, and [−1,2] for group 3.

- The smallest lower bound (in the example, −2) and the largest upper bound (in the example, 5) render a joint interval in the parameter space where the μ_js are most likely to be found.

- The normal prior for each μ_j can now be chosen such that μ_0 minus/plus one standard deviation is equal to the lower/upperbound of the joint interval, that is, $\mu_0 = (UB + LB)/2 = (5-2)/2 = 1.5$ and $\tau_0^2 = ((UB - LB)/2)^2 = ((5+2)/2)^2 = 3.5^2$, where UB and LB denote the upper and lower bounds of the joint interval, respectively. The resulting prior distribution allocates 95% of its density in the interval from −5.5 to 8.5, that is, according to the data it is rather unlikely that a population group mean is outside this interval.

Their prior for σ^2 is a scaled inverse chi-square distribution with degrees of freedom $\nu_0 = 1$ and scale parameter $\sigma_0^2 = s^2$, where s^2 is an estimate of σ^2 based on the data at hand.

This is an example of a default manner to specify the parameters of the prior distribution using a small amount of information from the data. The interested reader is referred to Kuiper and Hoijtink (2010) and Kuiper, Nederhoff, and Klugkist (Unpublished) for an elaboration and evaluation of this method. This method is implemented in the software package ConfirmatoryANOVA (Kuiper, Klugkist, and Hoijtink, 2010), which is illustrated in Chapter 8 and discussed in Section 9.2.1.

3.6.2.2   Prior Specification Method 2

Another approach is the procedure proposed by Mulder, Klugkist, and Hoijtink (2010), further elaborated by Mulder, Klugkist, van de Schoot, Meeus, Selfhout, and Hoijtink (2009), finalized in Mulder, Hoijtink, and de Leeuw (In Press), and evaluated for ANOVA models in van Wesel, Hoijtink, and Klugkist (In Press). This method is implemented in the software package BIEMS, which is illustrated in Appendix 3C and Chapters 4, 5, and 6. A short discussion of BIEMS can be found in Section 9.2.2. This approach is based on the work of Berger and Perrichi (1996, 2004) and Perez and Berger (2002) with respect to the use of training data for the specification of the parameters of prior distributions.

The minimum amount of data, the so-called minimal training sample (MTS) needed to be able to estimate a mean and a variance is two. Stated otherwise, an MTS consists of two observations randomly chosen from the data matrix. Let w denote a training sample, and let y_{iw} and $y_{i'w}$ denote the two observations chosen. Values for the parameters of the prior distribution of $\boldsymbol{\mu}$ are then obtained using

$$\mu_{0w} = (y_{iw} + y_{i'w})/2 \tag{3.18}$$

and

$$\tau^2_{0w} = \sigma^2_{0w}/2, \tag{3.19}$$

with

$$\sigma^2_{0w} = ((y_{iw} - \mu_0)^2 + (y_{i'w} - \mu_0)^2)/2. \tag{3.20}$$

To stress the dependence on the w-th MTS, the parameters μ_{0w} and τ^2_{0w} are indexed with w. The prior for σ^2 is Jeffreys prior, that is, $h(\sigma^2) \propto 1/\sigma^2$, which is obtained specifying a scaled inverse chi-square distribution with $\nu_0 = 0$ and an arbitrary value for σ^2_0.

The joint prior distribution of $\boldsymbol{\mu}$ and σ^2 can be rewritten as $h_w(\boldsymbol{\mu} \mid H_a)h(\sigma^2)$ because the first does and the latter does not depend on the training sample. As is elaborated in Chapter 10, the so constructed prior distribution is called the constrained posterior prior distribution (*CPP*). The name includes *constrained* because for $j = 1, \ldots, J$, each μ_j has the same the prior mean μ_0 and variance τ^2_0. The name includes *posterior* because a small amount of data is used to specify the prior distribution.

Multiple training samples $w = 1, \ldots, W$ can be used to avoid the dependence of inferences on one randomly chosen training sample. Multiple training samples result in W constrained posterior priors that can be averaged to render the expected constrained posterior prior (*ECPP*):

$$h(\boldsymbol{\mu}, \sigma^2 \mid H_a) = h(\boldsymbol{\mu} \mid H_a)h(\sigma^2) =$$

$$1/W \sum_{w=1}^{W} h_w(\boldsymbol{\mu} \mid H_a) \times h(\sigma^2). \tag{3.21}$$

It is computationally too intensive to use the expected constrained posterior prior for the computation of BF_{ma} because a sample from each of the W constrained posterior priors and the corresponding posteriors is needed to estimate c_m and f_m (Van Wesel, Hoijtink, and Klugkist, In Press). This can be avoided using an approximation based on a sample from (3.21), which is called the conjugate expected constrained posterior prior (*CECPP*):

- Obtain a sample of $\boldsymbol{\mu}$ of size T from each of the W posterior priors.

- Compute the mean $\boldsymbol{\mu}_0$ and covariance matrix $\boldsymbol{T}_0$ of the $T \times W$ sampled vectors $\boldsymbol{\mu}$. This renders $h(\boldsymbol{\mu} \mid H_a) \sim \mathcal{N}(\boldsymbol{\mu}_0, \boldsymbol{T}_0)$. This is a multivariate normal distribution of the form (3.1), that is, the prior means are equal, the prior variances are equal, and the prior covariances are equal.

3.6.3 Posterior Distribution

The posterior distribution of the unconstrained ANOVA model is

$$g(\boldsymbol{\mu}, \sigma^2 \mid \boldsymbol{y}, \boldsymbol{d}_1, \ldots, \boldsymbol{d}_J, H_a) = g(\boldsymbol{\mu}, \sigma^2 \mid \cdot, H_a) \propto$$

$$f(\boldsymbol{y}|\boldsymbol{d}_1, \ldots, \boldsymbol{d}_J, \boldsymbol{\mu}, \sigma^2)h(\boldsymbol{\mu} \mid H_a)h(\sigma^2). \tag{3.22}$$

The posterior distribution of an ANOVA model combined with hypothesis H_m is

$$g(\boldsymbol{\mu}, \sigma^2 \mid \cdot, H_m) = \frac{g(\boldsymbol{\mu}, \sigma^2 \mid \cdot, H_a)I_{\boldsymbol{\mu} \in H_m}}{\int g(\boldsymbol{\mu}, \sigma^2 \mid \cdot, H_a)I_{\boldsymbol{\mu} \in H_m}d\boldsymbol{\mu}, \sigma^2} =$$

$$1/f_m g(\boldsymbol{\mu}, \sigma^2 \mid \cdot, H_a)I_{\boldsymbol{\mu} \in H_m}, \tag{3.23}$$

where f_m is the proportion of the unconstrained posterior distribution in agreement with H_m.

3.7 Appendix 3B: Bayes Factor and Prior Sensitivity

3.7.1 Derivation of the Bayes Factor

The Bayes factor of H_m versus H_a is defined as the ratio of two marginal likelihoods (Kass and Raftery, 1995; Chib, 1995):

$$BF_{ma} = \frac{\int f(\boldsymbol{y} \mid \boldsymbol{d}_1, \ldots, \boldsymbol{d}_J, \boldsymbol{\mu}, \sigma^2)h(\boldsymbol{\mu}, \sigma^2|H_m)d\boldsymbol{\mu}, \sigma^2}{\int f(\boldsymbol{y} \mid \boldsymbol{d}_1, \ldots, \boldsymbol{d}_J, \boldsymbol{\mu}, \sigma^2)h(\boldsymbol{\mu}, \sigma^2|H_a)d\boldsymbol{\mu}, \sigma^2}. \tag{3.24}$$

Rewriting the marginal likelihoods as in Chib (1995), we obtain

$$BF_{ma} = \frac{f(\cdot)h(\boldsymbol{\mu}, \sigma^2|H_m)/g(\boldsymbol{\mu}, \sigma^2|\cdot, H_m)}{f(\cdot)h(\boldsymbol{\mu}, \sigma^2|H_a)/g(\boldsymbol{\mu}, \sigma^2|\cdot, H_a)} \tag{3.25}$$

for any value of $\boldsymbol{\mu}$ in agreement with the constraints of H_m. As can be seen, $f(\cdot)$ appears in both the numerator and the denominator and cancels. Furthermore, replacing $h(\boldsymbol{\mu}, \sigma^2|H_m)$ and $g(\boldsymbol{\mu}, \sigma^2|\cdot, H_m)$ by the outcome of (3.14) and (3.23), respectively, we obtain

$$BF_{ma} = f_m/c_m; \tag{3.26}$$

see, for example, Klugkist, Laudy, and Hoijtink (2005) and Klugkist and Hoijtink (2007).

3.7.2 Prior Sensitivity

3.7.2.1 The Prior for the Within Group Variance

In (3.14) in Appendix 3A is was shown that

$$c_m = \int h(\boldsymbol{\mu}, \sigma^2|H_a)I_{\boldsymbol{\mu} \in H_m}d\boldsymbol{\mu}, \sigma^2. \tag{3.27}$$

Because a priori $\boldsymbol{\mu}$ and σ^2 are independent, this can be written as

$$c_m = \int h(\boldsymbol{\mu} \mid H_a) h(\sigma^2) I_{\boldsymbol{\mu} \in H_m} d\boldsymbol{\mu}, \sigma^2$$

$$= \int_{\sigma^2} h(\sigma^2) d\sigma^2 \int_{\boldsymbol{\mu}} h(\boldsymbol{\mu} \mid H_a) I_{\boldsymbol{\mu} \in H_m} d\boldsymbol{\mu} = \int_{\boldsymbol{\mu}} h(\boldsymbol{\mu} \mid H_a) I_{\boldsymbol{\mu} \in H_m} d\boldsymbol{\mu}. \qquad (3.28)$$

Stated otherwise, the prior distribution of σ^2 does not influence c_m.

As is shown in (3.22), the posterior distribution does depend on the prior distribution of σ^2. However, if the data dominate the prior for σ^2, the posterior distribution (and consequently f_m) will hardly be affected by this prior. Stated otherwise, if $h(\sigma^2)$ is specified to be a vague distribution, it will hardly have an effect on the computation of f_m. Combined with the result from the previous paragraph this proves Property 3.1, that Bayes factors for informative hypotheses are virtually unaffected by the prior distribution for σ^2 as long as it is vague.

3.7.2.2   The Prior for the Means

As discussed in the context of two means using Figure 3.4, c_m for $H_1 : \mu_1 > \mu_2$ is independent of the choice of the prior distribution of $\boldsymbol{\mu}$ as long as $h(\mu_j) \sim \mathcal{N}(\mu_0, \tau_0^2)$ for $j = 1, 2$. Here we will prove that this results also holds for $H_1 : \mu_1 > \dots > \mu_J$ as long as $h(\boldsymbol{\mu}) \sim \mathcal{N}(\boldsymbol{\mu}_0, \boldsymbol{T}_0)$, where all the elements of $\boldsymbol{\mu}_0$ are equal to μ_0, each diagonal element of $\boldsymbol{T}_0$ is equal to τ_0^2, and each off-diagonal element is equal to τ_{00}.

Starting from (3.28) it can be shown for H_1 that

$$c_1 = \int_{\boldsymbol{\mu}} h(\boldsymbol{\mu} \mid H_a) I_{\boldsymbol{\mu} \in H_m} d\boldsymbol{\mu} =$$

$$Pr(\mu_1 > \dots > \mu_J \mid h(\boldsymbol{\mu} \mid H_a)) = 1/J!. \qquad (3.29)$$

Note that $\mu_1 > \dots > \mu_J$ is one of $J!$ ways in which J means can be ordered. Due to the symmetry of $h(\boldsymbol{\mu} \mid H_a)$, that is, equal prior means, variances, and covariances, it holds for any numbers $a_1 < \dots < a_J$ that $h(\mu_1 = a_1, \mu_2 = a_2, \dots, \mu_J = a_J | H_a) = \dots = h(\mu_1 = a_J, \mu_2 = a_{J-1}, \dots, \mu_J = a_1 | H_a)$. Stated otherwise, the prior density is identical for all $J!$ permutations by which $\boldsymbol{a} = [a_1, \dots, a_J]$ can be assigned to $\boldsymbol{\mu}$. This immediately implies that $c_1 = Pr(\mu_1 > \dots > \mu_J \mid h(\boldsymbol{\mu}|H_a)) = 1/J!$, independent of the choice of μ_0, τ_0^2, and τ_{00}.

As can be seen in (3.22), the posterior distribution (and consequently f_1) does depend on the prior distribution of $\boldsymbol{\mu}$. However, for $\tau_0^2 \to \infty$ and $\tau_0^2 \gg \tau_{00}$, the data will dominate the prior for $\boldsymbol{\mu}$ and this dependence will be negligible. Note that the prior specification methods presented in Sections 3.6.2.1 and 3.6.2.2 render vague prior distributions for $\boldsymbol{\mu}$. Stated otherwise, the corresponding posterior distributions will be virtually independent of these prior distributions.

Combined with the result from the previous paragraph, this proves Properties 3.2 and 3.6, which imply that BF_{a1} is virtually unaffected by the prior distribution for $\boldsymbol{\mu}$ as long as it is vague.

As illustrated using Figure 3.4, for $H_2 : |\mu_1 - \mu_2| < 1$, the complexity c_2 does depend on the prior distribution for $\boldsymbol{\mu}$. This implies that any Bayes factor involving H_2 also depends on the specification of $h(\boldsymbol{\mu}|H_a)$. The phenomenon underlying this dependence is the Lindley–Bartlett paradox (Lindley, 1957), which was originally formulated for hypotheses specified using exact equalities. Applied to the comparison of $H_2 : |\mu_1 - \mu_2| < 1$ and $H_a : \mu_1, \mu_2$, the Lindley–Bartlett paradox states: the larger τ_0^2, the stronger the support for H_2, that is, the larger BF_{2a}. As can be seen in Figure 3.4, the larger τ_0^2 the smaller c_2. Furthermore, for $\tau_0^2 \to \infty$, f_2 converges on a fixed data-dependent number f_{2*}. This results in $BF_{2a} =$

TABLE 3.2: data.txt

y_i	Group	y_i	Group
−.05	1	.29	2
.79	1	.06	2
.56	1	−.02	2
.47	1	−1.47	2
−.97	1	−1.38	2
−.88	1	.02	2
.52	1	1.91	2
2.41	1	.14	2
.64	1	.02	2
.52	1	−.00	2
.49	1	−.79	2
−.29	1	1.71	2
2.21	1	−.51	2
−.01	1	−.10	2
.39	1	−1.40	2
−.90	1	−1.15	2
−.65	1	1.31	2
1.81	1	1.04	2
1.54	1	.89	2
1.39	1	−.55	2

$f_{2*}/c_2 \to \infty$ if $\tau_0^2 \to \infty$. Stated otherwise, the larger τ_0^2 the larger the support in favor of H_2, irrespective of the information provided by the data! This explains, as was previously stated in Property 3.3, why the specification of τ_0^2 is important if hypotheses specified using equality or about equality constraints must be evaluated.

3.8   Appendix 3C: Using BIEMS for a Two Group ANOVA

Table 3.2 displays the data that were used for the Computation of BF_{1a} and BF_{0a} for $N = 20$ per group and $d = .5$ in Table 3.1. As is shown in Table 3.2, the data must be placed in a text file with the name data.txt. In the first column, the scores on the dependent variable are given and in the second column group membership is listed.

The commands needed to run BIEMS are listed in Table 3.3. These commands can either be placed in text files with the names input_BIEMS.txt and inequality_constraints.txt, or be entered using the BIEMS Windows user interface. The top panel of Table 3.3 contains the files needed to compute BF_{1a}; the bottom panel contains the files needed to compute BF_{0a}.

To compute BF_{1a}, only a few numbers must be entered (all the other numbers are set at default values): the number of dependent variables (#DV=1), the number of covariates (#cov=0), the number of time varying covariates (#Tcov=0), and the sample size (N=40). One line lower the the number of inequality constraints (#inequal=1) and equality constraints (#equal=0) are listed. In the bottom panel the inequality constraint of $H_1 : \mu_1 > \mu_2$ is listed. The string 1 -1 0 means states that $1 \times \mu_1 - 1 \times \mu_2 > 0$, which is equivalent to H_1. Note that the first two columns of the last line in the top panel represent the constraints matrix $\boldsymbol{R}_m$ and the last column the vector $\boldsymbol{r}_m$ (seen Chapter 2 for more examples of using these matrices and vectors to represent informative hypotheses).

To compute BF_{0a} (see the bottom panel of Table 3.3, only two numbers must be modified in `input_Biems.txt`: `#inequal=0` and `#equal=1`. This now indicates that the hypothesis of interest is specified using 1 equality constraint. The last line in the bottom panel states that this constraint is of the form $1 \times \mu_1 - 1 \times \mu_2 = 0$, which is equivalent to $H_0 : \mu_1 = \mu_2$.

Running `BIEMS` with the information in Tables 3.2 and 3.3 renders (among other things) the Bayes factor displayed in Table 3.1: $BF_{1a} = 1.85$ and $BF_{0a} = .98$, and, if the `BIEMS` `Windows` user interface is used, the corresponding posterior model probabilities: $PMP_1 = .65$ and $PMP_0 = .35$

TABLE 3.3: BIEMS Command Files

Computation of BF_{1a}					
`input_BIEMS.txt`					
Input 1:	#DV	#cov	#Tcov	N	iseed
	1	0	0	40	1111
Input 2:	#inequal	#equal	#restr		
	1	0	-1		
Input 3:	sample size	maxBF steps	scale		
	-1	-1	-1		
Input 4:	Z(DV)	Z(IV)			
	0	0			
Input 5:	Z(IV)				

`inequality_constraints.txt`	
$\boldsymbol{R}_m$	$\boldsymbol{r}_m$
1 -1	0

Computation of BF_{0a}					
`input_BIEMS.txt`					
Input 1:	#DV	#cov	#Tcov	N	iseed
	1	0	0	40	1111
Input 2:	#inequal	#equal	#restr		
	0	1	-1		
Input 3:	sample size	maxBF steps	scale		
	-1	-1	-1		
Input 4:	Z(DV)	Z(IV)			
	0	0			
Input 5:	Z(IV)				

`equality_constraints.txt`	
$\boldsymbol{S}_m$	$\boldsymbol{s}_m$
1 -1	0

Chapter 4

The J Group ANOVA Model

4.1 Introduction

Chapter 3 contained an introduction to Bayesian evaluation of informative hypotheses in the context of a two group ANOVA model. In this chapter the J group ANOVA model is used to illustrate the evaluation of informative hypotheses. Five situations are considered: hypotheses formulated using simple constraints, the situation where only one informative hypothesis is of interest, hypotheses formulated using constraints on combinations of means, hypotheses formulated using ordered means and effect sizes, and hypotheses formulated using about equality constraints. Three topics are discussed for each situation. First of all, what researchers themselves have to do: specify the hypotheses in which they are interested, and interpret the Bayes factors that result from executing software for the evaluation of

TABLE 4.1: Data Concerning Reduction in the Level of Aggression

	Group 1 Nothing	Group 2 Physical	Group 3 Behavioral	Group 4 Both
	1	1	4	7
	0	0	7	2
	0	0	1	3
	1	2	4	1
	−1	0	−1	6
	−2	1	2	3
	2	−1	5	7
	−3	2	0	3
	1	2	3	5
	−1	1	6	4
Sample Mean	−.20	.80	3.10	4.10
Standard Deviation	1.55	1.03	2.60	2.08

informative hypotheses. Second, it is illustrated how the software package BIEMS (Mulder, Hoijtink, and de Leeuw, In Press) can be used to obtain the Bayes factors. BIEMS is further discussed in Section 9.2.2. Third, the sensitivity of Bayes factors with respect to the prior distribution, and the performance of the Bayes factor are discussed.

4.2 Simple Constraints

4.2.1 Informative Hypotheses and Bayes Factors

In the J group ANOVA model, each of the J groups is characterized by a mean μ_j. The only other parameter is σ^2, which is the variance of the dependent variable within each of the J groups. Simple constraints consist of a set of comparisons of two means using the operators smaller than $<$, larger than $>$, equality $=$, and unconstrained ","; for example, "μ_1, μ_2" means that there is no constraint between μ_1 and μ_2.

In Section 1.2.2, a number of examples of informative hypotheses constructed using simple constraints were given for the example concerning reduction in aggression level. In this section the focus is on

$$H_0 : \mu_1 = \mu_2 = \mu_3 = \mu_4, \tag{4.1}$$

$$H_1 : \mu_1 < \mu_2 < \mu_3 < \mu_4, \tag{4.2}$$

and

$$H_a : \mu_1, \mu_2, \mu_3, \mu_4. \tag{4.3}$$

Remember that group 1 consisted of persons who received no training, group 2 contained persons receiving physical exercise, group 3 behavioral therapy, and group 4 both physical exercise and behavioral therapy.

Analysis of the data in Table 4.1 using the software package BIEMS rendered $BF_{0a} = .01$ and $BF_{1a} = 17.23$. As can be seen, there is an overwhelming amount of support for H_1; it is favored over H_a, which implies that the data support the constraints of H_1; and it received a much higher support than H_0.

TABLE 4.2: BIEMS Command Files for Section 4.2.2 "Simple Constraints"

Computation of BF_{1a} and BF_{11_c}

input_BIEMS.txt					
Input 1:	#DV	#cov	#Tcov	N	iseed
	1	0	0	40	1111
Input 2:	#inequal	#equal	#restr		
	3	0	-1		
Input 3:	sample size	maxBF steps	scale		
	-1	-1	-1		
Input 4:	Z(DV)	Z(IV)			
	0	0			
Input 5:	Z(IV)				

inequality_constraints.txt	
$\boldsymbol{R}_m$	$\boldsymbol{r}_m$
-1 1 0 0	0
0 -1 1 0	0
0 0 -1 1	0

As can be seen, Bayesian evaluation of informative hypotheses is rather straightforward: formulate the hypotheses, collect the data, and evaluate the hypotheses. It is important to note that Bayes factors are a relative measure of support, that is, they can be used to select the best of the hypotheses under consideration. However, the best of a set of hypotheses is not necessarily a hypothesis that gives a good description of the population from which the data were sampled. For this reason it is important to include H_a in the set of hypotheses under investigation. If the best of the hypotheses under investigation is also better than H_a, then the constraints used to formulate the hypothesis are supported by the data. This is such an important feature that it will be highlighted in a property:

Property 4.1: If the Bayes factor BF_{ma} is larger than 1, the constraints in H_m are supported by the data, that is, H_m provides a "fitting" description of the data.

If H_m is the best of a set of competing informative hypotheses, BF_{ma} can be used to verify whether H_m is only the best of a set of badly specified hypotheses, or, that it also has a good fit with the data.

An important question is whether these results imply that a researcher can conclude with confidence that the population from which the data were obtained can be described by H_1. This question has two aspects: to which degree are the results obtained sensitive to the specification of the prior distribution; and what is the performance of the Bayesian approach if (4.1) and (4.2) are evaluated in the context of an ANOVA model? Both aspects are discussed after an illustration of the use of BIEMS in the next section.

4.2.2 Implementation in BIEMS

To use BIEMS, the data displayed in Table 4.1 must be placed in a text file with the name data.txt. In the first column, the scores on the dependent variable are given and in the second column group membership is listed.

The commands needed to run BIEMS to compute BF_{1a} are listed in Table 4.2. These commands can either be placed in text files with the names input_BIEMS.txt and inequality_constraints.txt, or be entered using the BIEMS Windows user interface. In the top panel of Table 4.2, only four of the numbers must be entered (all the other numbers are set at default values): the number of dependent variables (#DV=1), the number of covariates (#cov=0), the number of time-varying covariates (#Tcov=0), and the sample size (N=40). One line lower, the the number of inequality constraints (#inequal=3) and equality constraints (#equal=0) are listed. In the bottom panel, the inequality constraints of $H_1 : \mu_1 < \mu_2 < \mu_3 < \mu_4$ are listed. The string -1 1 0 0 0 states that $-1 \times \mu_1 + 1 \times \mu_2 + 0 \times \mu_3 + 0 \times \mu_4 > 0$, which is equivalent to $\mu_1 < \mu_2$, that is, one of the constraints in H_1. Note that the first four columns in the bottom panel represent the constraints matrix $\boldsymbol{R}_m$ and the last column the vector $\boldsymbol{r}_m$ (seen Chapter 2 for more examples of using these matrices and vectors to represent informative hypotheses). Running BIEMS with the information in Tables 4.1 and 4.2 renders $BF_{1a} = 17.23$. Note that $BF_{0a} = .01$ is obtained by applying the following changes to the command files in Table 4.2: (#inequal=0), (#equal=3), and renaming inequality_constraints.txt to equality_constraints.txt, which changes $\boldsymbol{R}_m$ and $\boldsymbol{r}_m$ to $\boldsymbol{S}_m$ and $\boldsymbol{s}_m$, respectively.

The prior distribution used by BIEMS is a generalization of the prior distribution introduced in Chapter 3 for the two group ANOVA model. This generalization is also presented in Section 3.6.2 of Appendix 3A:

$$h(\boldsymbol{\mu} \mid H_a)h(\sigma^2) = \mathcal{N}(\boldsymbol{\mu}_0, \boldsymbol{T}_0)\text{Inv-}\chi^2(\nu_0, \sigma_0^2), \qquad (4.4)$$

in which each μ_j has the same prior mean μ_0 and variance τ_0^2, and in which the covariances are equal to τ_{00} for each pair of means. In BIEMS, the scaled inverse chi-square distribution is specified such that $h(\sigma^2) \propto 1/\sigma^2$. The specification $h(\boldsymbol{\mu} \mid H_a)$ rendered by BIEMS is $\mu_0 = 1.95$, $\tau_0^2 = 3.06$, and $\tau_{00} = .92$.

This prior distribution is neutral with respect to $H_1 : \mu_1 < \mu_2 < \mu_3 < \mu_4$. This hypothesis is one of $4! = 4 \times 3 \times 2 \times 1 = 24$ hypotheses in which four means are ordered from smallest to largest (another example is $H_{1'} : \mu_1 < \mu_3 < \mu_4 < \mu_2$). If the prior distribution is the same for each mean as it is in (4.4), then the proportion of the prior distribution in agreement with H_1 is 1/24. A proof of this statement can be found in Section 3.7.2.2 of Appendix 3B.

4.2.3   Sensitivity and Performance

4.2.3.1   Sensitivity

As elaborated before, the Bayes factor of H_m with respect to the unconstrained hypothesis H_a is equal to

$$BF_{ma} = f_m/c_m, \qquad (4.5)$$

where f_m and c_m are the proportions of the unconstrained prior and posterior, respectively, in agreement with the constraints of H_m. Properties 3.2 and 3.6 from Chapter 3 imply that both c_m and f_m (and thus BF_{ma}) are independent of the variance of the prior distribution τ_0^2 if H_m is specified using only inequality constraints and τ_0^2 is large.

For the example at hand (note that the sample size is relatively small: ten persons in each of four groups), the effect of τ_0^2 on BF_{1a} is displayed in Table 4.3 (computations are executed with a research version of ConfirmatoryANOVA, see Section 9.2.1). As can be seen, for all practical purposes the evaluation of BF_{1a} does not depend on τ_0^2: the data strongly support $H_1 : \mu_1 < \mu_2 < \mu_3 < \mu_4$. For larger sample sizes, the dependence of BF_{1a} on τ_0^2 will be even smaller.

Property 3.3 in Chapter 3 states that the size of τ_0^2 is important for the evaluation of informative hypotheses specified using equality or about equality constraints like $H_0 : \mu_1 = \mu_2 = \mu_3 = \mu_4$. This too is confirmed in Table 4.3: the larger τ_0^2, the larger BF_{0a}. However,

TABLE 4.3: BF_{0a} and BF_{1a} Computed by a Research Version of `ConfirmatoryANOVA` (see Section 9.2.1) using (4.4) with $\mu_0 = 1.95$, $\tau_{00} = 0$, $\nu_0 = 1$, and $\sigma_0^2 = 6.41$

τ_0^2	3	10	100	1,000	10,000	20,000
BF_{0a}	.00	.00	.02	.64	19.12	51.76
BF_{1a}	17.47	17.94	18.12	18.14	18.14	18.14

as can be seen in Table 4.3, for values of τ_0^2 in the range 3 to 100, the influence of τ_0^2 on BF_{0a} is negligible. Stated otherwise, for any *reasonable* value of τ_0^2, BF_{0a} is independent of τ_0^2. Note that a prior variance of 100 is already very unreasonable if the smallest observation is -3 and the largest observation is 7 as in Table 4.1.

4.2.3.2 Performance of the Bayes Factor

To illustrate the performance of the Bayes factor, data y are generated for four groups, each containing ten persons. Note again that Cohen's d is denoted by d if it concerns an effect size used to formulate an informative hypothesis, and by e if it is an effect size used to construct a data matrix or to specify a population. In each group, the variance of y is 1. In group 1, y has a mean of 0. The mean of y in groups 2, 3, and 4 depends on Cohen's c. If e is 0, the means in groups 2, 3, and 4 are equal to the mean in group 1. If e is .2, the means are 0, .2, .4, and .6, for groups 1, 2, 3, and 4, respectively; for $e = .5$ the means are 0, .5, 1.0, and 1.5, etc. The prior distribution used is (4.4).

In Table 4.4, BF_{0a} and BF_{1a} are displayed for different values of e. As can be seen, the Bayes factor performs adequately: if e increases, BF_{0a} decreases and BF_{1a} increases. Even with only ten persons per group, for $e = .5$ the support for H_1 (12.99) is convincingly larger than the support for H_0 (.17). This can be quantified by computing $BF_{10} = 12.99/.17 = 76.41$.

4.3 One Informative Hypothesis

4.3.1 Informative Hypothesis and Bayes Factors

The core of a research project may very well be one theory or expectation with respect to the population from which the data were sampled. Stated otherwise, researchers may be interested in the evaluation of one informative hypothesis without the desire to compare it to one or more competing hypotheses. The hypothesis of interest can be formulated using simple constraints, effect sizes, and constraints on combinations of means. Here the hypothesis

$$H_1 : \mu_1 < \mu_2 < \mu_3 < \mu_4, \tag{4.6}$$

TABLE 4.4: Performance of the Bayes Factor for Hypotheses Specified Using Simple Constraints

e	0	.2	.5	.8	1.0
BF_{0a}	9.83	4.91	.17	.01	.00
BF_{1a}	1	4.29	12.99	19.64	21.94

is used as an illustration. Note that this hypothesis was introduced in Chapter 1 in the context of the example concerning reduction in aggression, and was also used in the previous section. In this section it is compared to its complement H_{1_c}, that is, all possible orderings of four means excluding the ordering given in H_1.

Analysis of the data in Table 4.1 using the software package BIEMS rendered $f_1 = .721$ and $c_1 = .041$. Based on these results BIEMS rendered $BF_{11_c} = (f_1/c_1)/((1-f_1)/(1-c_1)) = (.721/.041)/(.279/.959) = 59.31$, which is convincing support in favor of H_1 over H_{1_c}.

4.3.2   Implementation in BIEMS, Sensitivity and Performance

Because the data and hypothesis of interest are the same as in the previous section, the implementation in BIEMS rendering BF_{11_c} displayed in Table 4.2 and the sensitivity are also the same and will not be further commented on. With respect to performance, one feature should be highlighted. The information used to compute BF_{1a} and BF_{11_c} is exactly the same. For three reasons the latter Bayes factor is to be preferred. First of all it has a simple interpretation: "H_1 or not H_1," which can also be paraphrased as "is it or is it not." Furthermore, comparison of H_1 with H_a is not logical because H_a encompasses H_1. Finally, as is illustrated by the example, $BF_{11_c} = 59.31$ is larger than $BF_{1a} = 17.23$. This implies that the evidence in favor or against H_1 if it is compared to H_{1_c} is always more pronounced (and thus easier to evaluate) than the evidence in favor or against H_1 if it is compared to H_a. The interested reader is referred to Van Rossum, van de Schoot, and Hoijtink (In Press) and Hoijtink (Unpublished) for further elaborations of the evaluation of one informative hypothesis.

4.4   Constraints on Combinations of Means

4.4.1   Informative Hypotheses and Bayes Factors

Hypotheses can also be specified using constraints on combinations of means. In Chapter 1 the following hypothesis was introduced for the reduction in aggression example:

$$H_1 : \begin{array}{l} (\mu_1 + \mu_3) < (\mu_2 + \mu_4) \\ (\mu_1 + \mu_2) < (\mu_3 + \mu_4) \\ (\mu_3 - \mu_1) < (\mu_4 - \mu_2). \end{array} \tag{4.7}$$

This hypothesis states that there is a positive effect of physical exercise, a positive effect of behavioral therapy, and a stronger effect of behavioral therapy if it is combined with physical exercise.

Another hypothesis that can be formulated is

$$H_2 : \quad (\mu_2 - \mu_1) < (\mu_3 - \mu_2) > (\mu_4 - \mu_3), \tag{4.8}$$

which states that the effect of physical exercise compared to no treatment is smaller than the effect of behavioral therapy compared to physical exercise, which in turn is larger than the effect of both compared to behavioral therapy.

Execution of BIEMS renders a Bayes factor for the comparison of H_1 with H_a of 3.76. Even with the third element of H_1 in (4.7) being in disagreement with the sample means displayed in Table 4.1 where the difference between the third and the first mean is equal to (instead of smaller than) the difference between the fourth and the second mean, there is

TABLE 4.5: BIEMS Command Files for Section 4.4.2 "Constraints on Combinations of Means"

Computation of BF_{2a}					
input_BIEMS.txt					
Input 1:	#DV	#cov	#Tcov	N	iseed
	1	0	0	40	1111
Input 2:	#inequal	#equal	#restr		
	2	0	3		
Input 3:	sample size	maxBF steps	scale		
	-1	-1	-1		
Input 4:	Z(DV)	Z(IV)			
	0	0			
Input 5:	Z(IV)				

inequality_constraints.txt	
R_m	r_m
1 -2 1 0	0
0 -1 2 -1	0

restriction_matrix.txt	
1 -1 0 0	0
0 1 -1 0	0
0 0 1 -1	0

still support for H_1 compared to H_a. The Bayes factor of H_1 with respect to H_{1_c} is equal to 6.21. As can once more be seen, comparing a hypothesis with its complement renders more extreme values of the Bayes factor.

The Bayes factor of H_2 with respect to H_a is 1.88. This may seem surprisingly small because the differences in sample means displayed in Table 4.1 are in agreement with $(\mu_2 - \mu_1) < (\mu_3 - \mu_2) > (\mu_4 - \mu_3)$. However, note that $c_1 = .12$ and $c_2 = .36$. This again illustrates that the Bayes factor is not only about fit, but also explicitly accounts for the complexity of a hypothesis when quantifying the support in the data for the hypotheses under investigation. The Bayes factor of H_1 with respect to H_2 is obtained via $BF_{12} = BF_{1a}/BF_{2a} = 3.76/1.88 = 2.00$.

4.4.2 Implementation in BIEMS

The commands needed to obtain BF_{2a} using BIEMS are listed in Table 4.5. Note that in the bottom panel, the string 1 -2 1 0 followed by 0 states that $1 \times \mu_1 - 2 \times \mu_2 + 1 \times \mu_3 + 0 \times \mu_4 > 0$, which is equivalent to the first element of H_2, that is, $(\mu_2 - \mu_1) < (\mu_3 - \mu_2)$. Running BIEMS with the information in Tables 4.1 and 4.5 renders $BF_{2a} = 1.88$. Changing #inequal=3 and placing restrictions in inequality_constraints.txt defining H_1 renders $BF_{1a} = 3.76$.

The default unconstrained prior distributions generated by BIEMS for H_1 and H_2 are not compatible, that is, are not the same. The importance of compatible prior distributions is stressed by the following property:

Property 4.2: The support in the data for two informative hypotheses H_m and $H_{m'}$ can only be quantified using $BF_{mm'} = BF_{ma}/BF_{m'a}$ if the unconstrained prior distributions corresponding to H_m and $H_{m'}$ are compatible, that is, identical.

The unconstrained prior distributions corresponding to H_m and $H_{m'}$ are compatible if H_m and $H_{m'}$ are the same if inequality constraints, about equality constraints and range constraints, are replaced by equality constraints. For example, (4.1) is identical to (4.2) if the inequality constraints are replaced by equality constraints. However, a similar maneuver cannot be used to change H_1 into H_2. In Section 9.2.2.1 the rules used to determine whether or not the unconstrained prior distributions of two competing hypotheses are compatible are elaborated.

In the case of incompatibility, the `BIEMS Windows` interface will automatically choose a prior distribution that is appropriate for all hypotheses under consideration (see Sections 4.5.2 and 5.8.1 for exceptions that will be announced by a warning). `BIEMS` can also manually be instructed to use compatible prior distributions. As can be seen in Table 4.5, `#restr=3`. This command states that `BIEMS` must generate a prior distribution using three restrictions that can be found in a command file with the name `restriction_matrix.txt`. This matrix can be found in the bottom panel of Table 4.5.

For the example at hand, `BIEMS` is instructed to give all parameters involved in the two hypotheses under consideration the same prior mean. Note that in `restriction_matrix.txt`, the line 1 -1 0 0 followed by 0 states that $1 \times \mu_{01} - 1 \times \mu_{02} + 0 \times \mu_{03} + 0 \times \mu_{04} = 0$, that is, the prior mean of μ_1 is equal to the prior mean of μ_2. The three restrictions in `restriction_matrix.txt` render an unconstrained prior distribution corresponding to the imaginary hypothesis $H_* : \mu_1 = \mu_2 = \mu_3 = \mu_4$. This prior distribution is the same as the prior distribution used in Section 4.2.2.

The restrictions in `restriction_matrix.txt` successfully render compatible prior distributions if H_* implies the constraints in H_m and $H_{m'}$ if all inequality, about equality, and range constraints have been replaced by equality constraints. If all the inequality constraints in H_2 are replaced by equality constraints, this renders $(\mu_2 - \mu_1) = (\mu_3 - \mu_2) = (\mu_4 - \mu_3)$. As can be seen, this constraint is implied by $H_* : \mu_1 = \mu_2 = \mu_3 = \mu_4$.

4.4.3   Performance

A small study was executed to evaluate the performance of the Bayes factor when evaluating informative hypotheses formulated using constraints on combinations of means. Table 4.6 presents the means $\bar{y}_j$ for $j = 1, \ldots, 4$ groups. In each group the within group variance is 1. The sample sizes used are $N = 10$ for each group.

The first row of Table 4.6 presents the results for the sample that is constructed to be in agreement with both H_1 and H_2. The resulting Bayes factors adequately show support for both hypotheses. Note that BF_{1a} is larger than BF_{2a} because the complexity of H_1 ($c_1 = .12$) is only 1/3 of the complexity of H_2 ($c_2 = .36$). Stated otherwise, H_1 is rewarded because it is a precise model with a good fit. The last three rows of Table 4.6 again show an adequate performance of the Bayes factor. Hypotheses in agreement with the sample at hand are supported, and hypotheses not in agreement are not supported.

TABLE 4.6: Evaluation of Hypotheses Formulated Using Constraints on Combinations of Means

True Hypotheses	$\bar{y}_1$	$\bar{y}_2$	$\bar{y}_3$	$\bar{y}_4$	f_1	c_1	BF_{1a}	f_2	c_2	BF_{2a}
H_1 & H_2	0	0	.5	.75	.37	.12	3.05	.53	.36	1.48
H_1	0	0	.5	1.5	.83	.12	6.75	.28	.36	.73
H_2	0	0	.5	0	.05	.12	.37	.67	.36	1.87
none	.75	.5	0	0	.01	.12	.10	.21	.36	.58

TABLE 4.7: BIEMS Command Files for Section 4.5.2 "Ordered Means with Effect Sizes"

Computation of BF_{1a} and BF_{11_c}

input_BIEMS.txt					
Input 1:	#DV	#cov	#Tcov	N	iseed
	1	0	0	40	1111
Input 2:	#inequal	#equal	#restr		
	3	0	-1		
Input 3:	sample size	maxBF steps	scale		
	-1	-1	-1		
Input 4:	Z(DV)	Z(IV)			
	0	0			
Input 5:	Z(IV)				

inequality_constraints.txt	
R_m	r_m
-1 1 0 0	.382
0 -1 1 0	.382
0 0 -1 1	.382

4.5 Ordered Means with Effect Sizes

4.5.1 Informative Hypotheses and Bayes Factors

Hypothesis (4.2) from the section on simple constraints can be reformulated such that effect sizes are included. Note that the within group standard deviation for the data in Table 4.1 is 1.91, and that $.2 \times 1.91$ denotes a small effect in terms of Cohen's d:

$$H_1 : \begin{array}{l} \mu_2 > \mu_1 + .2 \times 1.91 \\ \mu_3 > \mu_2 + .2 \times 1.91 \\ \mu_4 > \mu_3 + .2 \times 1.91 \end{array} , \tag{4.9}$$

which states that μ_2 is at least .2 standard deviations larger than μ_1, and the other two components have analogous interpretations.

Using BIEMS, the Bayes factor of H_1 with respect to H_a is 12.66. This implies that the data do support H_1. This is not surprising because the sample means in Table 4.1 are in agreement with the constraints in H_1. Note that the Bayes factor comparing H_1 with H_{1_c} is equal to $BF_{11_c} = 25.76$.

4.5.2 Implementation in BIEMS

The commands needed to run BIEMS to compute BF_{1a} are listed in Table 4.7. Note that the string -1 1 0 0 .382 in the bottom panel states that $-1 \times \mu_1 + 1 \times \mu_2 + 0 \times \mu_3 + 0 \times \mu_4 > .382$, which is equivalent to the first element of H_1. Executing BIEMS with these commands and the data in Table 4.1 rendered $BF_{1a} = 12.66$ and $BF_{11_c} = 25.76$.

The form of the prior distribution used is as in (4.4), but the specification of the parameters of the prior distribution is different from the other examples in this chapter: $\mu_0 = [1.36, 1.74, 2.12, 2.51]$, the diagonal elements of T are equal to $\tau_0^2 = 2.43$, and the off-diagonal elements are equal to $\tau_{00} = .84$. Note that the differences between the elements of μ_0 are $.2 \times 1.91 = .382$, and that their average is 1.95. This specification renders a prior distribution that is neutral with respect to H_1. Because there are 4! manners in which

TABLE 4.8: BF_{1a} Computed by a Research Version of `ConfirmatoryANOVA` (see Section 9.2.1) using (4.4) with $\boldsymbol{\mu}_0 = [1.36, 1.74, 2.12, 2.51]$, $\tau_{00} = 0$ $\nu_0 = 1$, and $\sigma_0^2 = 5.10$

τ_0^2	3	10	100	1,000	10,000	20,000
BF_{1a}	13.00	13.39	13.54	13.56	13.56	13.56

$\mu_1 + .2 \times 1.91$, $\mu_2 + .2 \times 1.91$, $\mu_3 + .2 \times 1.91$ and μ_4 can be ordered, c_1 should be (and is using the prior specification given) $1/4!$.

If $H_0 : \mu_1 = \mu_2 = \mu_3 = \mu_4$ is compared with (4.9), it can be seen that H_0 and H_1 are not the same (or can be rewritten to be the same) if all the inequality constraints are replaced by equality constraints, that is, H_0 and H_1 have incompatible unconstrained prior distributions. Note that H_0 and H_1 are inherently incompatible because there is no prior distribution corresponding to *both* hypotheses, that is, there is no imaginary hypothesis H_* which implies the constraints in both H_0 and H_1 (with all inequality constraints replaced by equality constraints).

A better null hypothesis corresponding to H_1 would be $H_{0'} : \mu_1 + 3 \times .382 = \mu_2 + 2 \times .382 = \mu_3 + .382 = \mu_4$. Furthermore, a researcher who wants to evaluate H_1 is probably interested in the size of the increase in the means. This implies that H_1 should be compared to the complementary hypothesis "not H_1." This would render an answer to the question "Are the differences between the means larger than $.2 \times 1.91 = .382$ or not?" This illustrates that, in general, the problem of inherently incompatible hypotheses can be solved using more appropriate competitors for the hypothesis of interest.

4.5.3   Sensitivity and Performance

4.5.3.1   Sensitivity

In Table 4.8 (computations are executed with a research version of `ConfirmatoryANOVA`, see Section 9.2.1), the sensitivity of $BF_{1,a}$ computed using the data in Table 4.1 with respect to the value of τ_0^2 is displayed for (4.9). As can be seen, $BF_{1,a}$ is not sensitive to the specification of τ_0^2.

4.5.3.2   Performance

In Table 4.9 the performance of BF_{1a} with

$$H_1 : \begin{array}{l} \mu_2 > \mu_1 + d \times s \\ \mu_3 > \mu_2 + d \times s \\ \mu_4 > \mu_3 + d \times s \end{array} \qquad (4.10)$$

is evaluated for two generated data sets that have sample means equal to 0, 0, 0, 0 labeled $e = 0$ and 0, .5, 1.0, 1.5 labeled $e = .5$, respectively. Note that e denotes differences between the sample means of e within group standard deviations s. In each data set, the sample size is 10 per group and the within group variance is 1 (consequently, the within group standard deviation $s = 1$).

TABLE 4.9: Performance of the Bayes Factor for a Hypothesis Specified Using Effect Sizes

$d \times s$ in H_1	0	.2	.5	.8	1.0
BF_{1a} for Data Generated Using $e = 0$	1.00	.17	.00	.00	.00
BF_{1a} for Data Generated Using $e = .5$	12.99	7.40	1.00	.02	.00

As can be seen, for $e = 0$ and $e = .5$, there is decreasing support for increasing effect sizes in the sample. Again, the behavior of the Bayes factor is satisfactory. If the difference between adjacent means is 0, it is clear that H_1 should not be supported by the data. If the difference between adjacent means is .5, values of $d \times s$ smaller than .5 should and values of $d \times s$ larger than .5 shouldn't be supported by the data. Note that the Bayes factor of 1.0 for, $e = 0$ and $d \times s = 0$, and $e = .5$ and $d \times s = .5$ implies that the ratio of fit to complexity is the same for H_1 and H_a. This is explained by the fact that for these situations the sample means are on the boundary between H_1 and H_a and that, consequently, the support for both hypotheses is the same.

4.6 About Equality Constraints

4.6.1 Informative Hypotheses and Bayes Factors

Hypothesis (4.1) can be changed to a hypothesis specified using effect sizes:

$$H_0 : \begin{array}{l} |\mu_1 - \mu_2| < .2 \times 1.91 \\ |\mu_2 - \mu_3| < .2 \times 1.91 \\ |\mu_3 - \mu_4| < .2 \times 1.91 \\ |\mu_1 - \mu_3| < .2 \times 1.91 \\ |\mu_1 - \mu_4| < .2 \times 1.91 \\ |\mu_2 - \mu_4| < .2 \times 1.91 \end{array} , \qquad (4.11)$$

which states that each pair of means differs by less than .2 within group standard deviations from each other. Note that for the data in Table 4.1, the within group standard deviation $s = 1.91$. In Cohen's terminology, this means that the pairwise effects are smaller than "small."

Executing **BIEMS** rendered a Bayes factor of H_0 with respect to H_a of .002. This implies that there is no support in the data for H_0 if both fit and complexity are accounted for. This is not surprising because the sample means in Table 4.1 are not in agreement with the constraints in H_0.

4.6.2 Implementation in BIEMS

The commands needed to run **BIEMS** to compute BF_{0a} are listed in Table 4.10. Note that the first two strings 1 -1 0 0 -.382 and -1 1 0 0 -.382 in the bottom panel state that $1 \times \mu_1 - 1 \times \mu_2 + 0 \times \mu_3 + 0 \times \mu_4 > -.382$ and, $-1 \times \mu_1 + 1 \times \mu_2 + 0 \times \mu_3 + 0 \times \mu_4 > -.382$, respectively. This is equivalent to the first element of H_0, that is, equivalent to $|\mu_1 - \mu_2| < .2 \times 1.91$. Executing **BIEMS** with these commands and the data in Table 4.1 rendered $BF_{0a} = .002$. The prior distribution used was the same as the prior distribution presented in Section 4.2 in which the focus was on simple constraints. This prior is in agreement with Property 3.4 from Chapter 3, it contains the same prior information for each of the means involved in H_0.

TABLE 4.10: BIEMS Command Files for Section 4.6.2 "About Equality Constraints"

Computation of BF_{0a}					
`input_BIEMS.txt`					
Input 1:	#DV	#cov	#Tcov	N	iseed
	1	0	0	40	1111
Input 2:	#inequal	#equal	#restr		
	12	0	-1		
Input 3:	sample size	maxBF steps	scale		
	-1	-1	-1		
Input 4:	Z(DV)	Z(IV)			
	0	0			
Input 5:	Z(IV)				

`inequality_constraints.txt`	
$\boldsymbol{R}_m$	$\boldsymbol{r}_m$
1 -1 0 0	-.382
-1 1 0 0	-.382
1 0 -1 0	-.382
-1 0 1 0	-.382
1 0 0 -1	-.382
-1 0 0 1	-.382
0 1 -1 0	-.382
0 -1 1 0	-.382
0 1 0 -1	-.382
0 -1 0 1	-.382
0 0 1 -1	-.382
0 0 -1 1	-.382

4.6.3   Sensitivity and Performance

4.6.3.1   Sensitivity

In Table 4.11 (computations are executed with a research version of `ConfirmatoryANOVA`, see Section 9.2.1), the sensitivity of $BF_{0,a}$ with respect to the value of τ_0^2 is displayed for H_0 as displayed in (4.11).

As can be seen, the sensitivity is the same as for $H_0 : \mu_1 = \mu_2 = \mu_3 = \mu_4$ discussed in Section 4.2, that is, the larger the prior variance τ_0^2, the larger the support for H_0. However, note that, here too, for reasonable values of τ_0^2, the evaluation of the Bayes factor renders conclusions that are, for all practical purposes, indistinguishable; there is no support in the data for H_0. Note that a prior variance of 100 is already very unreasonable if the smallest observation is -3 and the largest observation is 7 as in Table 4.1.

TABLE 4.11: BF_{0a} Computed by a Research Version of `ConfirmatoryANOVA` (see Section 9.2.1) using (4.4) with $\mu_0 = 1.95$, $\tau_{00} = 0$, $d \times s = .2 \times 1.91$, $\nu_0 = 1$, and $\sigma_0^2 = 6.41$

τ_0^2	3	10	100	1,000	10,000	20,000
BF_{0,a_0}	.0004	.003	.043	.67	66.66	200.00

TABLE 4.12: Performance of the Bayes Factor for a Null Hypothesis Specified Using One About Equality Constraint ($N = 20$ per Group)

Data Generated Using $e = 0$					
$d \times s$ in H_0	0	.5	1	2	3
BF_{0a}	2.72	1.98	1.28	1.02	1.0
BF_{00_c}		11.58	∞	∞	∞
f_0		.91	1.0	1.0	1.0
c_0		.46	.78	.99	1.0
Data Generated Using $e = .5$					
$d \times s$ in H_0	0	.5	1	2	3
BF_{0a}	.96	1.28	1.26	1.01	1.0
BF_{00_c}		1.68	9.69	∞	∞
f_0		.58	.97	1.0	1.0
c_0		.45	.77	.98	1.0

4.6.3.2 Performance of the Bayes Factor

The performance of the Bayes factor for hypotheses formulated using about equality constraints are discussed for the hypothesis $H_0 : |\mu_1 - \mu_2| < d \times s$, where $d \times s$ denotes a difference of d within group standard deviations. In Table 4.12, BF_{0a}, f_0, and c_0 are displayed. The panels in the table correspond to two effect sizes that were used to generate data. Using means of 0 and variances of 1 in both groups leads to $e = 0$. Keeping the variances at 1 and using means for groups 1 and 2, of 0 and .5, renders an effect size of $e = .5$.

First of all, some features will be highlighted:

- A $BF_{00_c} = \infty$ implies that $H_0 : |\mu_1 - \mu_2| < d \times s$ is infinitely more likely than $H_{0c} : |\mu_1 - \mu_2| > d \times s$. For example, if $e = 0$, the support for $H_0 : |\mu_1 - \mu_2| < 3$ is infinitely larger than the support for $H_{0c} : |\mu_1 - \mu_2| > 3$. For e equal to 0 and .5, it can be observed that $BF_{00_c} \to \infty$ for increasing $d \times s$. For the evaluations of inequality constrained hypotheses, BF_{00_c} is not very useful. It can only be used to disqualify H_0 if its value is smaller than 1.

- For $e = 0$, it can be observed that BF_{0a} decreases with increasing $d \times s$. Hypotheses with a small $d \times s$ are less complex (see c_0) than hypotheses with a large $d \times s$, and still have a good fit (see f_0). That is, the more "on target" a hypothesis is, the higher its Bayes factor.

- For $e = .5$, it can be observed that none of the values of $d \times s$ lead to a BF_{0a} convincingly indicating that the hypothesis at hand is an improvement over H_a. This is caused by the fact that hypotheses formulated with $d \times s > .5$ are not very informative, that is, c_0 is rather large.

These observations lead to three conclusions with respect to the evaluation of hypotheses formulated using about equality constraints:

- BF_{00_c} cannot be used for the evaluation of about equality constrained hypotheses.

- If $d \times s$ in H_0 is too large, the support in the data for H_0 will be very small. This is due to the large complexity of hypotheses with a large value of $d \times s$. In the context of about equality constraints, researchers are well advised to use values of $d \times s$ of .2 or smaller.

- If values for $d \times s$ larger than .2 are of interest, researchers are well advised to consider hypotheses formulated using range constraints like $H_0 : .2 < \mu_1 - \mu_2 < .5$ instead of hypotheses formulated using about equality constraints like $H_0 : |\mu_1 - \mu_2| < .5$. Range-constrained hypotheses can straightforwardly be evaluated using both BF_{0a} and BF_{00_c}.

4.7   Discussion

In this chapter, Bayesian evaluation of informative hypotheses as introduced in Chapter 3 was applied and evaluated in the context of the J group ANOVA model. It was shown that the Bayes factor showed satisfactory behavior for different sets of hypotheses. Two new properties where introduced: Property 4.1 stressed that the best of a set of competing hypotheses only gives an adequate description of the population of interest if $BF_{ma} > 1$ and Property 4.2 stressed that two informative hypotheses can only be compared if they have compatible prior distributions.

ANOVA models are based on the assumption of homogeneous within group variances (Miller, 1997). Null hypothesis significance testing is rather robust with respect to violations of this assumption if the sample sizes per group are about equal. The worst situation that can occur is a negative correlation between group sizes and within group variances. If the group sizes differ by more than a factor four, and the corresponding within group variances more than a factor ten, significance tests are no longer level α. Van Wesel, Hoijtink, and Klugkist (In Press) and Van Rossum, van de Schoot and Hoijtink (In Press) have evaluated the robustness of hypotheses evaluation using the Bayes factor with respect to violations of this assumption. The main conclusions are identical to those obtained for null hypothesis significance testing.

Chapter 3 discussed the foundations of Bayesian evaluation of informative hypotheses. In Chapter 4, Bayesian evaluation of informative hypotheses was illustrated and evaluated in the context of the J group ANOVA model. In Chapters 5 and 6, these illustrations and evaluations are continued for models belonging to the family of univariate and multivariate normal linear models, respectively. A new feature then enters the discussion: determination of the sample sizes needed in order to be able to reliably evaluate informative hypotheses using the Bayes factor.

Chapter 5

Sample Size Determination: AN(C)OVA and Multiple Regression

5.1   Introduction

In Chapter 3 the foundations of Bayesian evaluation of informative hypotheses were elaborated. Chapter 4 presented applications in the context of the J group ANOVA model, introduced two more properties (a reference value for interpretation of the Bayes factor and the requirement of compatible unconstrained prior distributions if the goal is to compare two informative hypotheses), discussed and illustrated sensitivity of the Bayes factor with respect to the specification of the unconstrained prior distribution, and showed how BIEMS can be used to obtain (among other things) Bayes factors.

This chapter starts with a discussion of sample size determination, that is, how large should the sample be to ensure that the data can be used to reliably distinguish among the informative hypotheses under consideration. Three approaches are available to researchers: a simple and fast approximate approach that is presented in this chapter; an approach based on the computation of error probabilities that is presented in the next chapter; and, inspired by the results presented in this and the next chapter, common sense. Subsequently, applications in the context of the univariate normal linear model are presented. ANOVA, ANCOVA and multiple regression models provide the context for the evaluation of informative hypotheses. In each context the three steps a researcher has to deal with when evaluating informative hypotheses are discussed: hypotheses formulation, sample size determination, and hypotheses evaluation.

In this chapter the focus is on the situation where a researcher has mainly one expectation that is translated into one informative hypothesis of interest. In the next chapter (in a context provided by the multivariate normal linear model), there is also attention for the evaluation of a series of nested informative hypotheses and a set of competing informative hypotheses.

5.2   Sample Size Determination

As is clear from the work of Cohen (1988, 1992) but see also Barker Bausell and Li (2006) and the work of Maxwell (2004), sample size determination is an important step when the goal is to evaluate hypotheses. After a discussion of classical power analysis, the main concepts involved will be applied to the evaluation of informative hypotheses.

5.2.1   Power Analysis

Consider the following hypotheses: the traditional null hypothesis

$$H_0 : \mu_1 = \mu_2, \qquad (5.1)$$

the traditional alternative hypothesis

$$H_a : \mu_1, \mu_2, \qquad (5.2)$$

and the one-sided hypothesis

$$H_m : \mu_1 > \mu_2. \qquad (5.3)$$

Note that these hypotheses refer to the means μ of a simple ANOVA model with two groups. The goal of null hypothesis significance testing is to determine whether or not H_0 can be

	H_0 is true	H_0 is not true
p-value > .05	Correct Decision: Do not reject H_0	H_0 is incorrectly not rejected The probability of this cell is called the error of the second kind
p-value < .05	Incorrect Rejection of H_0 The probability of this cell is called the alpha level	Correct Decision: Reject H_0 The probability of this cell is called the power

FIGURE 5.1: Alpha level and power explained.

rejected in favor of H_a. The criterion upon which the decision "reject or not" is based is the *p*-value.

Definition 5.1: The *p*-value is the probability of the observed data or data that are less in agreement with H_0 if H_0 is true.

Due to sampling fluctuations, the sample means for group 1 and group 2 computed for a data set randomly sampled from a population where H_0 is true will *not* be equal. The *p*-value is the probability of observing the sample means at hand, or, sample means that are even less in agreement with H_0, if H_0 is true. If the *p*-value is relatively large, a conventional benchmark is .05 (Rosnow and Rosenthal, 1989); the probability of the observed or more extreme sample means if H_0 is true is rather large, and H_0 is not rejected. If the *p*-value is smaller than .05, say .01, the probability is rather small and H_0 is rejected. The interested reader is referred to Berger and Sellke (1987), Howson and Urback (2006), Schervish (1996), Berger (2003), Wagenmakers (2007), and Van de Schoot, Hoijtink, Mulder, van Aken, Orobio de Castro, Meeus, and Romeijn (2011) for interpretations, evaluations, and elaborations of the *p*-value.

Loosely formulated, null hypothesis significance testing renders a transformation of the data into a *p*-value that is subsequently used to evaluate H_0. Of course the decision may be wrong. As can be seen in Figure 5.1, H_0 can be incorrectly rejected (the lower left-hand cell) or incorrectly not rejected (the upper right-hand cell). The probability of an incorrect rejection is called the alpha-level, which, by now almost conventional choice, is set at .05.

Definition 5.2: The alpha-level is the probability to incorrectly reject H_0.

If H_0 is not true, either H_0 is incorrectly not rejected (this so-called error of the second kind can be found in the upper right-hand corner of Figure 5.1) or H_0 is correctly rejected (power, see the lower right-hand corner of Figure 5.1). The focus of power analysis is to maximize the probability of the latter, and thereby to minimize the probability of the former.

Definition 5.3: Power is the probability to correctly reject H_0.

Like .05 is a conventional choice for the alpha-level, .80 is a conventional choice for the power (Cohen, 1992). The main goal of a power analysis is to determine the sample size needed to have a power of .80 to detect a certain effect size. A popular effect size measure for the example at hand is Cohen's d. Note that in this book, Cohen's d is denoted by e if it addresses effect sizes in samples and populations, and that Cohen's d is denoted by d if it is used to include an effect size in the formulation of informative hypotheses:

$$e = \frac{\mu_1 - \mu_2}{\sigma}, \qquad (5.4)$$

where σ is the within group standard deviation. It quantifies the difference between two means in terms of standard deviations. According to Cohen, .2 represents a small effect size, and .5 and .8, medium and large effect sizes, respectively. To be able to detect these effect sizes with a power of .80 using the t-test for the comparison of two independent means, samples of 393, 64, and 26 per group are needed (Cohen, 1992).

If H_0 is evaluated versus H_m instead of H_a, a one-sided t-test is obtained. If the effect is in the predicted direction, the alpha-level to which the resulting p-value should be compared is .10. In that case, to obtain a power of .80, samples of 310, 50, and 20 per group are needed (Cohen, 1992). Note that H_m is a very simple informative hypothesis. Compared to a noninformative alternative, with an informative alternative, smaller samples are needed to obtain the same power (for example, a sample of 50 instead of 64 per group to obtain a power of .80 to detect a medium-sized effect).

5.2.2   Unconditional and Conditional Error Probabilities

The alpha-level and the power are unconditional probabilities, that is, probabilities that can be computed before the data are observed.

Definition 5.4: Unconditional error probabilities are computed with respect to a hypothetical sequence of data matrices sampled from populations that correspond to the hypotheses under investigation. They are used to assess the performance of the procedure used for the evaluation of the hypotheses at hand.

Statements like "the probability to reject the null hypothesis if a data set is sampled from a population where the null is true is .05" are unconditional because they do not involve the data set at hand. Unconditional probabilities have their roots in classical statistics; they quantify properties of null hypothesis significance testing in an imaginary situation where data sets are repeatedly sampled from populations corresponding to the null and alternative hypothesis under investigation. Important determinants of error probabilities are sample size and effect size (Cohen, 1992; Barker Bausell, and Li, 2006): the larger the sample and/or effect size, the smaller the error probabilities.

The Bayesian approach has mostly focused on conditional error probabilities (Berger, Brown, and Wolpert, 1994; Berger, 2003).

Definition 5.5: Conditional error probabilities quantify the support in the observed data with respect to the hypotheses under investigation.

In fact, the $PMPs$ that were introduced in Chapter 3 can be interpreted as conditional error probabilities. Consider, for example, the situation where $BF_{0a} = 4$. Assuming that a priori H_0 and H_a are equally likely, this leads to $PMPs$ of .80 and .20 for H_0 and H_a, respectively.

These probabilities can be used in conditional statements like "given the data at hand the probability to incorrectly reject H_0 is .80 and the probability to incorrectly reject H_a is .20." However, as elaborated in Chapter 3, PMPs are not based on frequencies (like the probability of heads after flipping a coin), but on an evaluation of the fit and complexity of hypotheses. As such, they are logical probabilities (Carnap, 1950; Jeffreys, 1961; Jaynes, 2003), that is, numbers on a 0–1 scale that are an expression of degree of support. Note that $BF_{0a} = 1$ implies equal support for both hypotheses under investigation, that is, PMPs of .5 for both hypotheses. This explains the importance of the reference value 1 that is often used in the evaluation and interpretation of a Bayes factor (see also Property 4.1 in Chapter 4).

Both unconditional and conditional probabilities are appealing. The first can be used to determine the properties of a procedure if data sets are repeatedly sampled from populations specified by the hypotheses under investigation. The latter renders information with respect to the strength of the evidence (Royall, 1997) provided by the data set at hand. As is elaborated by Rubin (1984), Bayarri and Berger (2004), De Santis (2004), and Garcia-Donato and Chen (2005), both are of value, even if one has a preference for the Bayesian perspective.

Unconditional probabilities are useful because no one would like to endorse a procedure that chooses H_0 in, say, 60% of the cases if data sets are repeatedly sampled from H_a. This may very well happen, both for null hypothesis significance testing and hypothesis evaluation using the Bayes factor. If the sample sizes and/or the effect sizes are too small, H_0 may very well be the preferred hypothesis, even if data are repeatedly sampled from a population where H_a is true. Power analysis can be used to figure out what can and cannot be done with null hypothesis significance testing. As elaborated in this and the next chapter, the counterpart of power analysis in the Bayesian context is evaluation of the reliability of the Bayes factor.

Conditional probabilities are useful because, after a data set is observed, one would like a quantification of the certainty with which each of the hypotheses under investigation is endorsed. Where null hypothesis significance testing renders a dichotomous reject/not reject decision with respect to H_0, hypothesis evaluation using the Bayes factor renders the degree of support for each hypothesis under investigation.

In the next section a simple method to obtain information with respect to the reliability of the Bayes factor is introduced. At the end of this chapter this method is evaluated. In the next chapter a further elaboration of this method is presented. As will be shown, both the sample size and the effect size are important determinants of the reliability of the Bayes factor for the evaluation of informative hypotheses, that is, the larger the sample size and/or effect size, the higher the reliability of the Bayes factor.

5.2.3 Reliable Bayes Factors

As was shown in Section 5.2.1, with a sample of 64 per group, a medium effect size can reliably be detected using null hypothesis significance testing in the sense that the probability of incorrectly rejecting H_0 is .05, and the probability of incorrectly *not* rejecting H_0 is $1-.80=.20$. As argued in Section 5.2.2, sample size determination is also an important concept if the Bayes factor is used to evaluate hypotheses. The discussion of this topic begins with a definition of the reliability of the Bayes factor:

Definition 5.6: A set of informative hypotheses can reliably be evaluated if the Bayes factor provides convincing evidence in favor of a hypothesis if it is true, and convincing evidence against a hypothesis if it is not true.

TABLE 5.1: Guidelines for Reliable Bayes Factors I

Size of BF_{0a}	Interpretation
$1 - 3$	Not worth more than a bare mention
$3 - 20$	Positive
$20 - 150$	Strong
>150	Very strong

There are multiple answers to the question of what constitutes *convincing evidence*. Table 5.1 presents guidelines that can be found in the literature (Jeffreys, 1961; Kass and Raftery, 1995). Although these guidelines have an intuitive appeal, they are not well founded (De Santis (2004) and Garcia-Donato and Chen (2005)). What, for example, is the meaning of *Positive*? In terms of conditional error probabilities, $BF_{0a} = 19$ implies PMPs of .95 and .05 for H_0 and H_a, respectively. That is rather clear, but whether for the sample size at hand a value of the Bayes factor of 19 can be used to reliably distinguish H_0 from H_a still must be determined. Furthermore, why is 20 the value separating positive from strong? As Rosnow and Rosenthal (1989) formulated so eloquently in the context of null hypothesis significance testing, "Surely God loves the .06 as much as the .05." With the numbers presented in Table 5.1 another set of arbitrary guidelines is created. Therefore, in this book, these guidelines will not be used.

In this chapter the approach presented in Table 5.2 is used to get an impression of the sample size needed in order to obtain reliable Bayes factors. In the next chapter this approach is further elaborated. The Bayes factors in Table 5.2 are computed for data sets that correspond perfectly to the population at hand, that is, the effect size (Cohen's d denoted by e) in the sample is identical to the effect size in the population from which the data were sampled. Whereas the guidelines in Table 5.1 are rather general, the guidelines that can be derived from Table 5.2 are tailored to the situation at hand.

As can be seen in the top panel, where the Bayes factor of $H_0 : \mu_1 = \mu_2$ versus $H_a : \mu_1, \mu_2$ is presented, if the effect size is small, $e = .2$, neither a sample size of 20 nor 50 leads to a Bayes factor that is substantially different from the Bayes factor that is obtained is H_0 is true, that is, if $e = 0$. Furthermore, the size of the Bayes factor indicates support *for* instead of support *against* H_0; stated otherwise, using sample sizes of 20 and 50 does not render reliable Bayes factors for the comparison of H_0 with H_a if the effect size for H_a is small.

Looking at the results for a medium effect size of $e = .5$, it can be seen that with a sample size of 50 per group, the Bayes factor is in favor of H_a ($BF_{0a} = .25$ implies that the support in the data is 4 times larger for H_a than for H_0). Furthermore, the Bayes factor can clearly be distinguished from the Bayes factor that is obtained if H_0 is true (3.99). With a

TABLE 5.2: Reliable Bayes Factors: A Simple Example

Comparison of H_0 with H_a by means of BF_{0a}			
Sample Size	H_0 is true, $e = 0$	H_a is true, $e = .2$	H_a is true, $e = .5$
N=20 per group	2.78	2.35	.98
N=50 per group	3.99	2.56	.25
Comparison of H_0 with H_m by means of BF_{0m}			
Sample Size	H_0 is true, $e = 0$	H_m is true, $e = .2$	H_m is true, $e = .5$
N=20 per group	2.78	1.63	.53
N=50 per group	3.99	1.54	.13

TABLE 5.3: Guidelines for Reliable Bayes Factors II

	H_m is true			$H_{m'}$ is true	
$BF_{mm'}$	PMP_m	$PMP_{m'}$	$BF_{mm'}$	PMP_m	$PMP_{m'}$
1.01	.50	.50	.99	.50	.50
1.50	.60	.40	.80	.44	.56
4.00	.80	.20	.20	.17	.83

sample size of 50 per group and $e = .50$, the Bayes factor *can* be used to reliably distinguish between H_0 and H_a.

Note that for this simple situation, the results are in line with the results obtained using null hypothesis significance testing with an alpha level of .05 and a power of .80: to detect small effect sizes sample sizes of 393 per group (that is, much more than 50) are needed, and, to detect medium effect sizes, sample sizes of 64 per group (that is, about 50) are needed.

In the bottom panel of Table 5.2, H_a is replaced by H_m. As can be seen, this increases the evidence (and thus the reliability of the Bayes factor) against H_0 if it is not true.

Three principles were used to determine the reliability of the Bayes factor for the evaluation of the set of hypotheses under investigation:

- **Principle 5.1:** If hypothesis m is preferred (m may refer to an informative hypothesis, but also to the null, unconstrained, or complementary hypothesis), the Bayes factor for this hypothesis compared to hypothesis m' should be larger than 1.

The number 1 is a natural reference value for the evaluation of a $BF_{mm'}$. Values smaller than 1 point to H_m and values larger than 1 point to $H_{m'}$.

- **Principle 5.2:** $BF_{mm'}$ should be substantially different for the situations where H_m and $H_{m'}$ are true.

In Table 5.3 three different situations are displayed. Everybody will agree that the difference between Bayes factors of 1.01 (obtained if H_m is true) and .99 (obtained if $H_{m'}$ is true) is ignorable. This can also be seen from the corresponding PMPs. Most people will also agree that the difference between Bayes factors of 4 (if H_m is true) and .20 (if $H_{m'}$ is true) is rather large. This is also expressed by the corresponding PMPs of .80 in favor of H_m and .83 in favor of $H_{m'}$, respectively. However, researchers may very well have different opinions about whether Bayes factors of 1.50 and .80 are substantially different.

- **Principle 5.3:** To avoid new and arbitrary guidelines like the alpha-level of .05 in null hypothesis significance testing, a further specification of substantially different will not be given. This issue will return in the applications presented in the next sections.

As stated earlier, the approach used in this chapter can be used to get an impression of the sample sizes needed to obtain reliable Bayes factors for the hypotheses under investigation. The word *impression* is used to stress the fact that actual error probabilities are not computed. However, the information in tables like Table 5.2 give a good *impression* of the sample sizes needed to ensure that the error probabilities are acceptable for the hypotheses under investigation, without actually computing these error probabilities. In the next chapter a computationally rather intensive approach is introduced that does render the error probabilities.

In the next sections the approach described in this section is applied to the evaluation of informative hypotheses in the context of ANOVA, ANCOVA, and multiple regression. In Section 5.10 the approach used is evaluated.

5.3   ANOVA: Comparing Informative and Null Hypothesis

Sample size determination if the goal is to evaluate informative hypotheses in the context of ANOVA has previously been considered by Van Wesel, Hoijtink, and Klugkist (In Press), Van Rossum, van de Schoot, and Hoijtink (In Press) and Kuiper, Nederhoff, and Klugkist (Unpublised). The sample size tables presented in these papers can be useful in addition to the tables that are presented in this and the next section.

5.3.1   Hypotheses

The research context is the same as in Section 1.1: the dependent variable is a person's decrease in aggression level between week 1 (intake) and week 8 (end of training); and the data consist of persons in need of anger management training that will randomly been assigned to one of four groups: 1. no training, 2. physical exercise, 3. behavioral therapy, and 4. physical exercise and behavioral therapy. The first step a researcher must execute is translation of his expectations in one or more informative hypotheses. The informative hypothesis considered in this section is $H_1 : \mu_1 < \mu_2 < \mu_3 < \mu_4$. It will be evaluated against the null hypothesis $H_0 : \mu_1 = \mu_2 = \mu_3 = \mu_4$.

5.3.2   Sample Size Determination

A researcher has the option to obtain an idea about the sample sizes needed to be able to make a reliable distinction between H_1 and H_0 using the Bayes factor. Sample size determination for the evaluation of informative hypotheses, like power analysis in hypothesis testing, is based on a hypothetical line of argument. The following ingredients are needed:

- The hypotheses under investigation. In this section, H_1 will be compared to H_0.

- A range of sample sizes in which the researcher is interested. Here, four sample sizes will be considered: 10, 20, 40, and 80.

- A range of effect sizes in which the researcher is interested. Here, six effect sizes will be considered: the difference between two adjacent means will in terms of Cohen's d (denoted by e) be 0, .1, .2, .3, .4, or .5.

With this information a table can be generated from which the sample size needed per group to reliably distinguish among the hypotheses under investigation can be determined. The following three-step procedure was used to compute the entries in Table 5.4:

- Step 1: Provide the means in each group and the within group standard deviation in the population. For the first entry in Table 5.4, the means for groups 1 to 4 were 0, 0, 0, 0 and the within group standard deviation 1, that is, e for adjacent groups is 0. For the last entry on the first line of Table 5.4, the means were 0, .5, 1.0, 1.5 also with a within group standard deviation of 1, that is, $e = .5$. Analogous choices were made for all other entries in the table.

- Step 2: Create a data set with the sample size per group as indicated in the first column of Table 5.4 that has sample means and within group standard deviation that are exactly equal to their population counterparts. Such a data set is the best representation of the population for the sample size at hand; stated otherwise, it is the expected data set for the sample size at hand. Such data sets can be generated

TABLE 5.4: ANOVA Sample Size Determination, Table Entries are BF_{10}

Sample Size per Group	$e = 0$	$e = .1$	$e = .2$	$e = .3$	$e = .4$	$e = .5$
10	.11	.27	.79	2.34	11.29	49.92
20	.04	.21	1.33	14.22	254.83	2302.50
40	.02	.19	5.66	559.00	>1,000	>1,000
80	.01	.20	>1,000	>1,000	>1,000	>1,000

using the program `GenMVLData` with the option `exact=1`. In Table 5.22 in Appendix 5B the command file `input.txt` needed to run `GenMVLData` to obtain the data set for $N = 20$ and $e = .2$ in Table 5.4 is discussed. In Section 9.2.4, further information with respect to `GenMVLData` can be found. Note that `GenMVLData` can be executed from the `BIEMS Windows` user interface.

- Step 3: Use `BIEMS` to Compute BF_{10} for each entry in Table 5.4. The command files needed to obtain BF_{10} from BIEMS were presented in Section 4.2.2.

With respect to Principles 5.1 and 5.2, the following can be observed in Table 5.4:

- Even for sample sizes as small as 10 per group, the Bayes factor is smaller than 1 and clearly supports H_0 if the data correspond to a population where $e = 0$ for adjacent means.

- For sample sizes as small as 20 per group, the Bayes factor clearly supports H_1 if the data correspond to a population where $e \geq .3$ for adjacent means, that is, the Bayes factor is larger than 1 and clearly different from the corresponding Bayes factor for $e = 0$.

- The larger the sample sizes per group, the smaller the effects that can reliably be detected using the Bayes factor.

Note that here and in the remainder of this chapter the author has tried to formulate conclusions such that most if not all scientists agree without giving a formal definition of "substantially different" (see Principle 5.3). Nevertheless, each scientist is well advised to obtain his own evaluation of Table 5.4. There will probably be some that would change the second bullet above to "For sample sizes as small as 10 per group, ..." Similar considerations hold for the other tables presented in this chapter.

Depending on the effect size expected by a researcher in his population of interest, Table 5.4 can be used to get an indication of the sample size needed in order to be able to reliably evaluate the hypotheses of interest using the Bayes factor. In our example, the effect sizes were expected to be relatively large. Consequently, it was decided to include 10 persons in each group. The data are displayed in Table 5.5.

5.3.3 Hypotheses Evaluation Using BIEMS

The previous section was all about unconditional error probabilities, that is, how large should the sample size be in order to obtain reliable Bayes factors. In this section the Bayes factor and $PMPs$ obtained for the data set at hand have to be interpreted, which is all about conditional error probabilities.

As was shown in Section 4.2.1, the Bayes factor of H_1 versus H_0 obtained using the data in Table 5.5 was $BF_{10} = BF_{1a}/BF_{0a} = 17.23/.01 = 1723$ which constitutes overwhelming evidence in favor of H_1, that is, the support in the data for H_1 is much larger than for

TABLE 5.5: Data with Respect to the Decrease in Aggression Level

	Group 1 Nothing	Group 2 Physical	Group 3 Behavioral	Group 4 Both
	1	1	4	7
	0	0	7	2
	0	0	1	3
	1	2	4	1
	−1	0	−1	6
	−2	1	2	3
	2	−1	5	7
	−3	2	0	3
	1	2	3	5
	−1	1	6	4
Sample Mean	−.20	.80	3.10	4.10
Standard Deviation	1.55	1.03	2.60	2.08

H_0. This means that it can be concluded that the combination of physical exercise and behavioral therapy has a better effect than the use of only one therapy form, which in turn has a better effect than no therapy at all. Note that the sample means displayed in Table 5.5 are firmly in agreement with H_1. Details of the computation with BIEMS and the specification of the prior distribution can be found in Section 4.2.2.

In other applications, smaller values of the Bayes factor may be observed. If, for example, $BF_{10} = 9$, this implies that the $PMPs$ of H_1 and H_0 are .9 and .1, respectively. Still this is substantial evidence in favor of H_1. But what if $BF_{10} = 2$? In this case, the $PMPs$ for H_1 and H_0 are .67 and .33, respectively. In this situation there is not a very clear distinction between both hypotheses. Stated otherwise, the data at hand do not have a clear preference for either of the hypotheses under consideration. To avoid the creation of new and arbitrary "rules" like ".05" as the alpha-level in null hypothesis significance testing, guidelines and benchmarks like those in Table 5.1 that can be used to evaluate Bayes factors and $PMPs$ will not be given in this book.

5.4   ANOVA: Comparing Informative Hypothesis and Complement

5.4.1   Hypotheses

The research context is the same as in the previous section. However, in this section the goal is to compare H_1 with its complement H_{1_c} : not H_1.

5.4.2   Sample Size Determination

The ingredients for sample size determination are

- The hypotheses H_1 and H_{1_c}.

- Sample sizes of 10, 20, 40, 80, and 160.

- None, one small, one large, two small, and three small deviations from the order of the means hypothesized in H_1.

TABLE 5.6: ANOVA Sample Size Determination, Table Entries are BF_{11_c}

Sample Size per Group	None	One Small	One Large	Two	Three
10	4.57	4.38	1.05	.86	.11
20	9.33	3.84	.63	.52	.02
40	20.44	3.18	.18	.16	.00
80	51.05	2.22	.01	.01	.00
160	175.98	.79	.00	.00	.00

With this information a table can be generated from which the sample size needed per group to reliably distinguish among the hypotheses under investigation can be determined. The following three-step procedure was used to compute the entries in Table 5.6:

- Step 1: In each population the within group standard deviation is 1, in the "None" population $e = .2$ between adjacent means, and there are no violations of the order specified in H_1: the population means are 0, .2, .4, and .6. In the "One Small" population there was 1 small deviation of the order specified in H_1: .2, 0, .4, and .6. In the "One Large" population there was 1 large deviation of the order specified: .5, 0, .4, .6. In the "Two" and "Three" violations populations, the means were .4, .2, 0, .6 and .6, .4, .2, 0, respectively.

- Step 2: Use `GenMVLData` with the option `exact=1` to create a data set with the sample size per group as indicated in the first column of Table 5.6 that has sample means and within group standard deviation that are exactly equal to their population counterparts. The command file needed is the same as that used in Section 5.3.2.

- Step 3: Use `BIEMS` to compute BF_{11_c} for each entry in Table 5.6. The command files needed were presented in Section 4.2.2.

In line with Principles 5.1 and 5.2 the following can be observed in Table 5.6.

- A sample size of 10 per group is sufficient to find support for H_1 if it is true (even if the effect sizes are small like in the population at hand).

- A sample size of 20 per group is sufficient to find support for H_{1_c} (for one large, two and three violations) that is substantially different from the corresponding support for H_1.

- If there is one small $d = .2$ violation the situation is more complicated. This is due to the fact that with one small violation H_1 is roughly correct and small sample sizes are not sufficient to detect small violations. For sample sizes of 80 and smaller, it will often incorrectly be concluded that H_1 is supported by the data because Bayes factors larger than 1.0 are observed. However, note that the Bayes factors in the first column (H_1 is true) are (with the exception of $N = 10$) substantially larger than the Bayes factors in the second column (H_{1_c} is true).

5.4.3 Hypotheses Evaluation Using BIEMS

As was previously shown in Section 4.3.1 using the data displayed in Table 5.5, $BF_{11_c} = 59.31$, that is, H_1 is strongly supported by the data. The corresponding $PMPs$ are .98 for H_1 and .02 for H_{1_c}, respectively, that is, in terms of conditional error probabilities there is convincing evidence in favor of H_1. This implies that the ordering of treatment effects

TABLE 5.7: ANCOVA Sample Size Determination, Table Entries are BF_{10}

Sample Size per Group	$e = 0$	$e = .1$	$e = .2$	$e = .3$	$e = .4$	$e = .5$
10	.15	.44	1.56	7.43	38.97	243.00
20	.07	.46	5.44	124.36	>1,000	>1,000
40	.02	.52	64.90	>1,000	>1,000	>1,000
80	.01	1.64	>1,000	>1,000	>1,000	>1,000

in H_1 is the best of all possible orderings of treatment effects. As can be seen in Table 5.6, a sample size of 10 per group implies that three and maybe two small violations of the predicted order can be detected, but does not ensure detection of one small or large violation. However, because 59.31 is more than 10 times larger than the Bayes factors listed under "None" and "One Small" violation in Table 5.6, it is very likely that H_1 is correct with an effect size larger than the effect size used to construct the data set with no violations and a sample size of 10 in Table 5.6.

5.5   ANCOVA

5.5.1   Hypotheses

Section 5.3 presented an ANOVA example in which persons were randomly assigned to one of four groups. In this subsection this example will be elaborated via the inclusion of the covariate *Age* of each person in the analysis. This changes the ANOVA to an ANCOVA. See Section 1.3 for an introduction and discussion of this ANCOVA example. As in Section 5.3, the goal is to compare $H_1 : \mu_1 < \mu_2 < \mu_3 < \mu_4$ with $H_0 : \mu_1 = \mu_2 = \mu_3 = \mu_4$. However, note that due to the inclusion of the covariate *Age*, the μs in both hypotheses now address adjusted means.

5.5.2   Sample Size Determination

The following ingredients are needed:

- The hypotheses under investigation. In this section H_1 will be compared to H_0.

- A range of sample sizes in which the researcher is interested. Here four sample sizes will be considered: 10, 20, 40 and 80.

- A range of effect sizes in which the researcher is interested. Here six effect sizes will be considered: the difference between two adjacent adjusted means will in terms of Cohen's d (denoted by e) be 0, .1, .2, .3, .4 or .5.

The following three-step procedure was used to compute the entries in Table 5.7:

- Step 1: The sample means and standard deviations of the dependent variable are exactly as in Table 5.4; to stress this, Table 5.7 is labeled analogously. The difference is that a covariate has been added that has a mean of 0 and a variance of 1 in each group, and within each group correlates .6 with the dependent variable. Compared to the ANOVA, this reduces the within group variance to $1 - .6^2 = .64$.

TABLE 5.8: Data Suited for ANCOVA

	Group 1 Nothing			Group 2 Physical			Group 3 Behavioral			Group 4 Both	
y	x	Group	y	x	Group	y	x	Group	y	x	Group
0	18	1	3	23	2	4	21	3	6	21	4
0	20	1	1	24	2	3	22	3	5	22	4
0	21	1	1	19	2	4	23	3	5	23	4
1	22	1	1	20	2	5	25	3	6	25	4
2	23	1	2	21	2	5	26	3	6	24	4
1	24	1	1	18	2	5	27	3	6	23	4
0	19	1	1	20	2	4	23	3	6	26	4
0	21	1	2	22	2	3	21	3	6	27	4
0	20	1	3	23	2	5	22	3	6	24	4
1	22	1	1	21	2	4	25	3	5	23	4

- Step 2: Use `GenMVLData` with the option `exact=1` to create a data set with the sample size per group as indicated in the first column of Table 5.7 that has sample statistics that are exactly equal to their population counterparts: adjusted means, within group standard deviation, regression coefficient relating the covariate to the dependent variable, mean of the covariate, and variance of the covariate. In Table 5.23 in Appendix 5B the command file `input.txt` needed to run `GenMVLData` to obtain the data set for $N = 20$ and $e = .2$ in Table 5.7 is presented.

- Step 3: Use `BIEMS` to Compute BF_{10} for each entry in Table 5.7. The commands needed to compute BF_{10} are presented in the next section.

Because the covariate has a mean of 0 and a variance of 1 in each group, it is not included to account for differences among the groups with respect to the covariate. However, inclusion of a covariate, even in randomized experiments, may reduce the sample sizes needed to reliably evaluate the hypotheses of interest. This can be seen comparing the entries of Table 5.7 with Table 5.4. The Bayes factors obtained for the ANCOVA for $e \geq .2$ are larger than the corresponding Bayes factors obtained for the ANOVA. Using Principles 5.1 and 5.2 it can be concluded that a sample size of $N = 10$ is sufficient to reliably detect effect sizes of $e = .3$ and larger. A sample size of $N = 20$ is sufficient to reliably detect effect sizes of $e = .2$ and larger.

5.5.3   Hypotheses Evaluation Using `BIEMS`

Because it is expected that the effect sizes will be $e = .3$ or larger, the sample contained 10 persons per group. The data are displayed in Table 5.8. Note that x denotes the covariate age and y the reduction in aggression and "Group" group number. For analysis with `BIEMS` the data must be placed in a file with the name `data.txt` with reduction in agression in the first column, followed by age and finally group number. Table 5.9 displays the estimates of the adjusted means, the regression coefficient relating decrease in aggression level to standardized age , and their standard errors obtained using ANCOVA as implemented in SPSS (http://www.spss.com/). Note that the covariate is standardized to ensure that the μs are adjusted means. As can be seen, the estimates of the adjusted means are in agreement with H_1.

Analysis of the data using `BIEMS` rendered $BF_{10} = BF_{1a}/BF0a = 24.00/.01 = 2400$, which constitutes overwhelming evidence in favor of H_1. That is, after correction for the covariate age, the combination of both physical exercise and behavioral therapy leads to a

TABLE 5.9: Estimates of Adjusted Means and Regression Coefficient for the Data in Table 5.8

	μ_1	μ_2	μ_3	μ_4	β
Estimate	.81	1.89	3.93	5.36	.53
Standard Error	.19	.19	.19	.20	.11

larger reduction in aggression than each treatment separately, which in turn has a better effect than no treatment at all. Note that the corresponding PMPs for H_1 and H_0 are 1.00 and .00, respectively.

The command files needed to analyze the data in Table 5.8 with **BIEMS** can be found in Table 5.10. Note that the number of covariates **#cov=1** to announce the presence of a covariate. Note also that for the computation of BF_{1a}, the number of inequality constraints **#inequal=3**, while for the computation of BF_{0a}, the number of equality constraints **#equal=3**. Note furthermore that **Z(IV)=1** denotes that the covariate has to be standardized. Note finally that the command file containing the constraints is called **inequality_constraints.txt** if it contains inequality constraints and **equality_constraints.txt** if contains equality constraints. The string **-1 1 0 0 0** combined with 0 states that $-1 \times \mu_1 + 1 \times \mu_2 + 0 \times \mu_3 + 0 \times \mu_4 + 0 \times \beta > 0$, that is, compared to the corresponding ANOVA input (see Section 4.2.2), the regression coefficient β, which relates *Age* to the decrease in aggression, has been added. The command files in Table 5.10 render BF_{1a} and BF_{0a}. The Bayes factor of interest is subsequently obtained as $BF_{10} = BF_{1a}/BF_{0a}$.

5.6   Signed Regression Coefficients: Informative versus Null

5.6.1   Hypotheses

The example to be used in this section was introduced in Section 1.4. A multiple regression model is used to predict standardized *Income* from standardized *IQ* and standardized *SES*. Note that both the dependent variable and the predictors are standardized to ensure comparability of the regression coefficients of both predictors:

$$Z(Income_i) = \beta_0 + \beta_1 Z(IQ_i) + \beta_2 Z(SES_i) + \epsilon_i. \tag{5.5}$$

In this section the informative hypothesis of interest is

$$H_1 : \begin{array}{l} \beta_1 > 0 \\ \beta_2 > 0 \end{array}. \tag{5.6}$$

It will be compared to the classical null hypothesis:

$$H_0 : \begin{array}{l} \beta_1 = 0 \\ \beta_2 = 0 \end{array}. \tag{5.7}$$

TABLE 5.10: BIEMS Command Files for Hypotheses Evaluation in the Context of AN-COVA

Computation of BF_{1a}					
input_BIEMS.txt					
Input 1:	#DV	#cov	#Tcov	N	iseed
	1	1	0	40	1111
Input 2:	#inequal	#equal	#restr		
	3	0	-1		
Input 3:	sample size	maxBF steps	scale		
	-1	-1	-1		
Input 4:	Z(DV)	Z(IV)			
	0	1			
Input 5:	Z(IV)				

inequality_constraints.txt					
R_m					r_m
-1	1	0	0	0	0
0	-1	1	0	0	0
0	0	-1	1	0	0

Computation of BF_{0a}					
input_BIEMS.txt					
Input 1:	#DV	#cov	#Tcov	N	iseed
	1	1	0	40	1111
Input 2:	#inequal	#equal	#restr		
	0	3	-1		
Input 3:	sample size	maxBF steps	scale		
	-1	-1	-1		
Input 4:	Z(DV)	Z(IV)			
	0	1			
Input 5:	Z(IV)				

equality_constraints.txt					
S_m					s_m
-1	1	0	0	0	0
0	-1	1	0	0	0
0	0	-1	1	0	0

5.6.2 Sample Size Determination

The following ingredients are needed:

- The hypotheses under investigation. In this section, H_1 will be compared to H_0.

- A range of sample sizes in which the researcher is interested. Here, four sample sizes will be considered: 12, 20, 40, and 80.

- Populations representing the range of effect sizes in which the researcher is interested must be constructed. Assuming that both predictors have the same regression coefficient, proportions of variance explained R^2 of 0, .1, .3, and .5 will be considered. Because the correlation between both predictors influences the proportion of variance explained, three situations will be distinguished: $\rho_{IQ,SES}$ equal to 0, .3, and .5.

TABLE 5.11: Sample Size Determination for Signed Regression Coefficients (Table Entries are BF_{10})

Sample Size	$R^2 = 0$ $\beta_1 = \beta_2 = 0$	$\rho_{IQ,SES} = 0$ $R^2 = .1$ $\beta_1 = \beta_2 = .22$	$R^2 = .3$ $\beta_1 = \beta_2 = .39$	$R^2 = .5$ $\beta_1 = \beta_2 = .50$
12	.15	.51	1.35	4.62
20	.06	.34	3.50	30.00
40	.04	.64	48.88	>1,000
80	.02	2.48	>1,000	>1,000

Sample Size	$R^2 = 0$ $\beta_1 = \beta_2 = 0$	$\rho_{IQ,SES} = .3$ $R^2 = .1$ $\beta_1 = \beta_2 = .20$	$R^2 = .3$ $\beta_1 = \beta_2 = .34$	$R^2 = .5$ $\beta_1 = \beta_2 = .44$
12	.13	.40	1.25	6.32
20	.07	.37	3.92	47.75
40	.03	.71	55.57	>1,000
80	.01	2.83	>1,000	>1,000

Sample Size	$R^2 = 0$ $\beta_1 = \beta_2 = 0$	$\rho_{IQ,SES} = .5$ $R^2 = .1$ $\beta_1 = \beta_2 = .18$	$R^2 = .3$ $\beta_1 = \beta_2 = .32$	$R^2 = .5$ $\beta_1 = \beta_2 = .41$
12	.07	.35	1.36	4.87
20	.06	.27	2.25	62.00
40	.02	.49	47.00	393.00
80	.01	2.22	400.00	>1,000

The following three-step procedure was used to compute the entries in Table 5.11:

- Step 1: In each population, both the dependent and predictor variables have mean 0 and variance 1. The intercept β_0 has the value 0. The values of β_1 and β_2 (both have the same value) rendering the desired effect sizes can be found in Table 5.11.

- Step 2: Use `GenMVLData` with the option `exact=1` to create a data set with the sample size per group as indicated in the first column of Table 5.11 that has sample statistics that are exactly equal to their population counterparts: regression coefficients, residual variance, and means and covariance matrix of the two predictors. In Table 5.24 in Appendix 5B the command file `input.txt` needed to run `GenMVLData` to obtain the data set for $N = 20$, $\rho_{IQ,SES} = .3$ and $\beta_0 = 0, \beta_1 = .20, \beta_2 = .20$ in Table 5.11 is discussed.

- Step 3: Use `BIEMS` to compute BF_{10} for each entry in Table 5.11. The commands needed to compute BF_{10} are presented in the next section.

If the goal is to compare H_1 with H_0, with respect to Principles 5.1 and 5.2 the following can be observed in Table 5.11:

- If H_0 is true the BF_{10} is much smaller that 1, that is, there is convincing support for H_0.

- If the sample size is 20 or larger and the R^2 is .3 or larger, BF_{10} is substantially larger than 1 and quite different from its counterpart when H_0 is true.

In summary, the Bayes factor can reliably be used to compare H_1 and H_0 if the sample size is larger than 20 for populations where R^2 is at least .3. As can also be seen in Table 5.11, sample sizes of at least 80 are needed to obtain reliable Bayes factors for effect sizes of .1.

TABLE 5.12: Data Suited for Multiple Regression

Income	IQ	SES	Group	Income	IQ	SES	Group
51.09	79	1	1	66.12	97	4	1
40.88	82	3	1	66.81	100	5	1
41.29	83	4	1	69.09	100	4	1
57.74	85	4	1	80.32	100	5	1
57.01	87	4	1	64.47	110	4	1
41.18	88	3	1	65.20	116	6	1
58.24	95	2	1	80.31	117	7	1
52.19	95	3	1	83.93	120	5	1
59.54	96	5	1	67.84	121	4	1
67.40	96	3	1	64.46	123	5	1

5.6.3   Hypotheses Evaluation Using BIEMS

The results obtained in the previous section suggest that a sample of 20 persons should suffice for a reliable comparison of H_1 with H_0. Note that analysis with BIEMS requires that the data are placed in a text file with the name data.txt containing *Income* in the first column, subsequently *IQ* and *SES*, and finally a column consisting of ones indicating that all persons belong to the same group. Analysis of the data in Table 5.12 rendered $BF_{10} = BF_{1a}/BF_{0a} = 3.43/.01 = 343$. The corresponding *PMPs* for H_1 and H_0 are 1.00 and .00, respectively, which constitutes convincing evidence in favor of H_1. It can be concluded that larger values of both *IQ* and *SES* correspond with larger values of *Income*. Note also that the regression coefficients in Table 5.13 estimated with multiple regression as implemented in SPSS (http://www.spss.com/) are firmly in agreement with H_1.

Table 5.14 presents the BIEMS command files needed to compute BF_{1a} and BF_{0a}, which may subsequently be combined to render $BF_{10} = BF_{1a}/BF_{0a}$. Note that #cov=2 states that there are two predictors; Z(DV)=1 states that the dependent variable is standardized; Z(IV)=1 states that the predictors are standardized; and the string 0 1 0 followed by 0 in the top panel states that $0 \times \beta_0 + 1 \times \beta_1 + 0 \times \beta_2 > 0$, which is equivalent to the first component of H_1. The same string in the bottom panel states that $0 \times \beta_0 + 1 \times \beta_1 + 0 \times \beta_2 = 0$, which is equivalent to the first component of H_0.

5.7   Signed Regression Coefficients: Informative and Complement

5.7.1   Hypotheses

The main difference between this section and the previous section is that H_1 will be compared to H_{1_c} instead of H_0.

TABLE 5.13: Estimates of Standardized Regression Coefficients for the Data in Table 5.12

	β_0	β_1	β_2
Estimate	.00	.58	.32
Standard Error	.16	.21	.21

TABLE 5.14: BIEMS Command Files for the Evaluation of Signed Hypotheses in the Context of Multiple Regression

Computation of BF_{1a} and BF_{11_c}					
input_BIEMS.txt					
Input 1:	#DV	#cov	#Tcov	N	iseed
	1	2	0	20	1111
Input 2:	#inequal	#equal	#restr		
	2	0	-1		
Input 3:	sample size	maxBF steps	scale		
	-1	-1	-1		
Input 4:	Z(DV)	Z(IV)			
	1	1			
Input 5:	Z(IV)				

inequality_constraints.txt	
$\boldsymbol{R}_m$	r_m
0 1 0	0
0 0 1	0

Computation of BF_{0a}					
input_BIEMS.txt					
Input 1:	#DV	#cov	#Tcov	N	iseed
	1	2	0	20	1111
Input 2:	#inequal	#equal	#restr		
	0	2	-1		
Input 3:	sample size	maxBF steps	scale		
	-1	-1	-1		
Input 4:	Z(DV)	Z(IV)			
	1	1			
Input 5:	Z(IV)				

equality_constraints.txt	
$\boldsymbol{S}_m$	s_m
0 1 0	0
0 0 1	0

5.7.2 Sample Size Determination

Table 5.15 is constructed analogously to Table 5.11 from the previous section. The only difference is that it displays BF_{11_c} instead of BF_{10}.

Guided by Principles 5.1 and 5.2, the following feature can be observed in Table 5.15: sample sizes of 20 and larger provide substantial evidence in favor of H_1 if the effect size R^2 is at least .1, that is, the Bayes factor is larger than 1 and substantially different from the corresponding Bayes factor for $R^2 = 0$. Note that $R^2 = 0$ corresponds to $\beta_1 = \beta_2 = 0$, that is, a combination of regression coefficients on the boundary between H_1 and H_{1_c}, that is, the combination of regression coefficients closest to H_1 that no longer belongs to H_1. Stated otherwise, H_1 does not hold if $R^2 = 0$, and as can be seen in Table 5.15, in this situation H_1 is not supported by the data.

In many situations the value of the Bayes factor for least favorable populations on the boundary between two hypotheses is 1, that is, neither hypothesis is preferred over the other. As can be seen in Table 5.15, these observations are not a general rule, if $\rho_{IQ,SES} = 0$, BF_{11_c} is indeed equal to 1; but if $\rho_{IQ,SES} > 0$, BF_{11_c} is smaller than 1. At first sight this may seem to be an undesirable feature of the Bayes factor. However, the interested reader is

TABLE 5.15: Sample Size Determination for Signed Regression Coefficients (Table Entries are BF_{11_c})

Sample Size	$\rho_{IQ,SES} = 0$ $R^2 = 0$ $\beta_1 = \beta_2 = 0$	$R^2 = .1$ $\beta_1 = \beta_2 = .22$	$R^2 = .3$ $\beta_1 = \beta_2 = .39$	$R^2 = .5$ $\beta_1 = \beta_2 = .50$
12	1	3.37	8.59	20.21
20	1	5.39	24.96	157.95
40	1	14.16	263.95	>1,000
80	1	62.17	>1,000	>1,000

Sample Size	$\rho_{IQ,SES} = .3$ $R^2 = 0$ $\beta_1 = \beta_2 = 0$	$R^2 = .1$ $\beta_1 = \beta_2 = .20$	$R^2 = .3$ $\beta_1 = \beta_2 = .34$	$R^2 = .5$ $\beta_1 = \beta_2 = .44$
12	.78	2.51	6.32	15.48
20	.78	4.01	17.77	71.45
40	.78	9.66	96.91	>1,000
80	.78	31.80	>1,000	>1,000

Sample Size	$\rho_{IQ,SES} = .5$ $R^2 = 0$ $\beta_1 = \beta_2 = 0$	$R^2 = .1$ $\beta_1 = \beta_2 = .18$	$R^2 = .3$ $\beta_1 = \beta_2 = .32$	$R^2 = .5$ $\beta_1 = \beta_2 = .41$
12	.63	1.97	5.04	10.56
20	.63	2.86	11.01	43.01
40	.63	6.27	45.78	442.56
80	.63	16.41	564.99	>1,000

referred to Appendix 5A where it is explained why the Bayes factor for boundary values can be smaller (or larger) than 1.

5.7.3 Hypotheses Evaluation Using BIEMS

Because $N = 20$, BF_{11_c} can reliably be used to compare H_1 with H_{1_c}. Analysis of the data in Table 5.12 rendered $BF_{11_c} = 18.28$. Note that the corresponding PMPs are .95 and .05, respectively, which constitute convincing evidence in favor of H_1. It can be concluded that the relations between IQ and SES on the one side and *Income* on the other side are positive, and not otherwise. BIEMS renders BF_{11_c} if the command files in the top panel of Table 5.14 are used.

5.8 Signed Regression Coefficients: Including Effect Sizes

5.8.1 Hypotheses

If the dependent variable and the predictors are standardized, the regression coefficients are standardized, that is, are numbers on a scale from -1 to 1. The latter facilitates their interpretation and thus the addition of effect sizes to an informative hypothesis. In this section the focus is on informative hypotheses of the form

$$H_1 : \begin{array}{l} \beta_1 > d \\ \beta_2 > d \end{array} , \tag{5.8}$$

where d denotes the smallest effect size a researcher considers to be interesting. Note that in this book, d denotes an effect size used to formulate an informative hypothesis. In the context of analysis of variance, d represents Cohen's d. In this section, d represents the smallest value of a standardized regression coefficient considered to be interesting.

If a hypothesis is formulated including an effect size, there are several options for the hypothesis to which it can be compared, among which are

$$H_0 : \begin{matrix} \beta_1 = 0 \\ \beta_2 = 0 \end{matrix} \;, \tag{5.9}$$

$$H_0 : \begin{matrix} \beta_1 = d \\ \beta_2 = d \end{matrix} \;, \tag{5.10}$$

and,

$$H_{1_c} : \text{not } H_1. \tag{5.11}$$

The interpretation of a comparison of (5.8) with (5.11) is straightforward: are both coefficients larger than d or not? However, if the desire is to compare (5.8) with a point null hypothesis, should it be (5.9) or (5.10)? Both are valid alternatives; which is preferred depends on the research question that must be answered. However, the prior distributions for β_0, β_1, and β_2 under (5.8) and (5.9) are inherently incompatible (see Section 4.4.2 and Property 4.2 in Chapter 4 for a discussion of compatibility): if the inequality constraints in H_1 are replaced by equality constraints, the result is not identical to H_0 in (5.9); and, there is not an imaginary hypothesis H_* whose constraints imply the constraints in H_0 in (5.9) and H_1 (with the inequality constraints replaced by equality constraints). This implies that with the methodology presented in this book (5.8) can be compared to (5.10) but not to (5.9).

Equation (1.46) from Appendix 1A can be used to determine the proportion of variance that will at least be explained if β_1 and β_2 at least have the value d. For the example at hand, d is chosen such that (5.8) and (5.10) state that R^2 is and larger than and equal to .1, respectively. Given the observed correlation of .653 between the scores on IQ and SES as displayed in Table 5.12, this is achieved using $d = .174$, resulting in

$$H_1 : R^2 > .1, \text{ that is } \begin{matrix} \beta_1 > .174 \\ \beta_2 > .174 \end{matrix} \;, \tag{5.12}$$

and

$$H_0 : R^2 = .1, \text{ that is } \begin{matrix} \beta_1 = .174 \\ \beta_2 = .174 \end{matrix} \;. \tag{5.13}$$

5.8.2   Sample Size Determination

Table 5.16 is the counterpart of Table 5.11 for the evaluation of (5.12) versus (5.13). Again guided by Principles 5.1 and 5.2, the following features can be observed for BF_{10}:

- If R^2 is equal to the reference value of .1 (implied by the choice of $d = .174$), there is convincing evidence in favor of $H_0 : R^2 = .1$.

- If R^2 is smaller than the reference value of .1, the evidence is strongly in favor of H_0, that is, the least incorrect of both hypotheses under investigation is preferred.

- If R^2 is .5 or larger and N is 40 or larger, there is convincing evidence in favor of $H_1 : R^2 > .1$ over $H_0 : R^2 = .1$.

TABLE 5.16: Sample Size Determination for Signed Regression Coefficients Including Effect Sizes (Table Entries are BF_{10})

Sample Size	$R^2 = 0$ $\beta_1 = \beta_2 = 0$	$R^2 = .1$ $\beta_1 = \beta_2 = .22$	$R^2 = .3$ $\beta_1 = \beta_2 = .39$	$R^2 = .5$ $\beta_1 = \beta_2 = .50$
	$\rho_{IQ,SES} = 0$			
12	.08	.16	.41	1.05
20	.02	.07	.24	1.31
40	.01	.04	.49	14.92
80	.00	.02	1.07	400.00

Sample Size	$R^2 = 0$ $\beta_1 = \beta_2 = 0$	$R^2 = .1$ $\beta_1 = \beta_2 = .20$	$R^2 = .3$ $\beta_1 = \beta_2 = .34$	$R^2 = .5$ $\beta_1 = \beta_2 = .44$
	$\rho_{IQ,SES} = .3$			
12	.06	.13	.32	.89
20	.02	.05	.19	1.04
40	.01	.03	.33	12.43
80	.00	.01	.72	396.00

Sample Size	$R^2 = 0$ $\beta_1 = \beta_2 = 0$	$R^2 = .1$ $\beta_1 = \beta_2 = .18$	$R^2 = .3$ $\beta_1 = \beta_2 = .32$	$R^2 = .5$ $\beta_1 = \beta_2 = .41$
	$\rho_{IQ,SES} = .5$			
12	.05	.10	.28	.77
20	.01	.04	.16	1.00
40	.00	.02	.33	12.96
80	.00	.01	.93	400.00

Table 5.17 is the counterpart of Table 5.15 for the evaluation of H_1 versus its complement H_{1_c}. The following features can be observed:

- If $R^2 = .1$, which is on the boundary between H_1 and H_{1_c}, there is no convincing evidence in favor of either hypothesis.

- If $R^2 < .1$, there is convincing evidence in favor of H_{1_c}.

- IF R^2 is larger or equal to .3, there is convincing evidence in favor of H_1 for sample sizes of 20 and larger.

5.8.3 Hypotheses Evaluation Using BIEMS

The data in Table 5.12 contain the responses of 20 persons to *Income, IQ*, and *SES*. This sample size is not sufficient to reliably compare H_1 and H_0 for the effect sizes displayed in Table 5.16. However, it is sufficient for a reliable comparison of H_1 and H_{1_c} for effect sizes R^2 equal to or larger than .3.

BIEMS renders $BF_{10} = BF_{1a}/BF_{0a} = 2.35/.76 = 3.09$, that is, the support in the data is about 3 times larger for H_1 than for H_0. Looking at Table 5.16 in the panel for $\rho_{IQ,SES} = .5$ (because in the observed data this correlation is .653) at the line $N = 20$, it can be seen that 3.09 is a rather large value. Stated otherwise, although $N = 20$ is not sufficient to reliably compare H_1 and H_0 for the effect sizes displayed in Table 5.16, it may very well be that the effect size in the population from which the data were sampled is larger than $R^2 = .5$. Note that in the data $R^2 = .56$, that is, $BF_{10} = 3.09$ can be considered positive evidence in favor of $H_1 : R^2 > .1$. Note also that this conclusion is supported by the regression coefficients displayed in Table 5.13. Note furthermore that $BF_{11_c} = 4.27$. The corresponding $PMPs$ are .81 and .19 for H_1 and H_{1_c}, respectively. These results imply that the data favor H_1,

TABLE 5.17: Sample Size Determination for Signed Regression Coefficients Including Effect Sizes (Table Entries are BF_{11_c})

Sample Size	$R^2 = 0$ $\beta_1 = \beta_2 = 0$	$R^2 = .1$ $\beta_1 = \beta_2 = .22$	$R^2 = .3$ $\beta_1 = \beta_2 = .39$	$R^2 = .5$ $\beta_1 = \beta_2 = .50$
		$\rho_{IQ,SES} = 0$		
12	.30	1.02	2.80	6.49
20	.15	1.02	4.44	16.83
40	.03	1.03	9.90	101.67
80	.00	1.02	31.87	2928.51

Sample Size	$R^2 = 0$ $\beta_1 = \beta_2 = 0$	$R^2 = .1$ $\beta_1 = \beta_2 = .20$	$R^2 = .3$ $\beta_1 = \beta_2 = .34$	$R^2 = .5$ $\beta_1 = \beta_2 = .44$
		$\rho_{IQ,SES} = .3$		
12	.22	.81	2.10	4.75
20	.10	.76	3.05	9.99
40	.02	.78	6.00	41.64
80	.00	.76	14.11	406.23

Sample Size	$R^2 = 0$ $\beta_1 = \beta_2 = 0$	$R^2 = .1$ $\beta_1 = \beta_2 = .18$	$R^2 = .3$ $\beta_1 = \beta_2 = .32$	$R^2 = .5$ $\beta_1 = \beta_2 = .41$
		$\rho_{IQ,SES} = .5$		
12	.19	.67	1.80	3.85
20	.09	.63	2.53	7.32
40	.02	.62	4.70	26.80
80	.00	.61	11.11	171.84

but that H_{1_c} cannot yet be ruled out completely. All this leads to the conclusion that it is rather likely but not yet certain that IQ and SES explain at least 10% of the variation in *Income*.

To obtain BF_{1a} and BF_{11_c}, the command file `input_BIEMS.txt` is the same as in the top panel of Table 5.14. However, the command file `inequality_constraints.txt` changes and can be found in the top panel of Table 5.18. To obtain BF_{0a}, the command file `input_BIEMS.txt` is the same as in the bottom panel of Table 5.14. However, the command file `equality_constraints.txt` corresponding to H_0 can be found in the bottom panel of Table 5.18. Note that the string 0 1 0 followed by .174 in the top panel states that $0 \times \beta_0 + 1 \times \beta_1 + 0 \times \beta_2 > .174$. The same string in the bottom panel states that $0 \times \beta_0 + 1 \times \beta_1 + 0 \times \beta_2 = .174$.

TABLE 5.18: BIEMS Command Files for the Evaluation of Signed Hypotheses Including Effect Sizes in the Context of Multiple Regression

Computation of BF_{1a} and BF_{11_c}	
`inequality_constraints.txt`	
R_m	r_m
0 1 0	.174
0 0 1	.174
Computation of BF_{0a}	
`equality_constraints.txt`	
S_m	s_m
0 1 0	.174
0 0 1	.174

5.9 Comparing Regression Coefficients

5.9.1 Hypotheses

In this subsection we continue with the prediction of *Income* from *IQ* and *SES* using multiple regression (see (5.5)), however the informative hypothesis of interest now is

$$H_1 : \beta_1 > \beta_2, \tag{5.14}$$

that is, *IQ* is a stronger predictor of *Income* that *SES*. The competing hypotheses will again be a null hypothesis, $H_0 : \beta_1 = \beta_2$, and the complementary hypothesis H_{1_c} : not H_1.

5.9.2 Sample Size Determination

The following ingredients are needed:

- The hypotheses under investigation. In this section H_1 will be compared to H_0 and H_{1_c}.

- A range of sample sizes in which the researcher is interested. Here, four sample sizes will be considered: 12, 20, 40, and 80.

- Proportions of variance explained R^2 of .32 and .5 and correlations between both predictors $\rho_{IQ,SES}$ equal to 0 and .5. will be considered. For each of three combinations of R^2 and $\rho_{IQ,SES}$, the difference between β_1 and β_2 varies from 0 via approximately .2 to approximately .3.

The following three-step procedure was used to compute the entries in Tables 5.19 and 5.20:

- Step 1: In each population, both the dependent and predictor variables have mean 0 and variance 1. The intercept β_0 has the value 0. The values of β_1 and β_2 rendering the desired effect sizes can be found in Tables 5.19 and 5.20.

- Step 2: Use `GenMVLData` to create a data set with the sample size per group as indicated in the first column of Tables 5.19 and 5.20. Using the option `exact=1` generates a data set with sample statistics corresponding exactly to the generating population, that is, sample regression coefficients, sample residual variance, and sample means and sample covariance matrix of the two predictors are exactly equal to their population counterparts. Table 5.24 in Appendix 5B presents the command file used to generate such data with `GenMVLData`.

- Step 3: Use `BIEMS` to compute BF_{10} for each entry in Table 5.19 and to compute BF_{11_c} for each entry in Table 5.20. The `BIEMS` commands needed for these computations are presented in the next section.

With respect to Principles 5.1 and 5.2, the comparison of H_1 with H_0 in Table 5.19 has the following features:

- If H_0 is true, for both sizes of the sample and R^2, BF_{10} is smaller than 1.0, that is, the Bayes factor correctly prefers H_0.

- If H_1 is true with a difference between β_1 and β_2 of about .2, sample sizes larger than 80 are needed to obtain a Bayes factor that is substantially larger than 1.0, that is, a Bayes factor that prefers H_1 over H_0.

TABLE 5.19: Sample Size Determination for the Comparison of Regression Coefficients (Table Entries are BF_{10})

	$R^2 = .32$ and $\rho_{IQ,SES} = 0$		
Sample Size	$\beta_1 = .4, \beta_2 = .4$	$\beta_1 = .48, \beta_2 = .3$	$\beta_1 = .53, \beta_2 = .2$
12	.55	.61	.88
20	.42	.62	1.02
40	.26	.62	1.90
80	.18	.80	6.35
	$R^2 = .50$ and $\rho_{IQ,SES} = 0$		
Sample Size	$\beta_1 = .5, \beta_2 = .5$	$\beta_1 = .58, \beta_2 = .4$	$\beta_1 = .64, \beta_2 = .3$
12	.50	.83	.95
20	.37	.65	1.36
40	.28	.86	3.34
80	.19	1.08	19.9
	$R^2 = .50$ and $\rho_{IQ,SES} = .5$		
Sample Size	$\beta_1 = .41, \beta_2 = .41$	$\beta_1 = .51, \beta_2 = .3$	$\beta_1 = .59, \beta_2 = .2$
12	.56	.98	1.19
20	.53	.90	1.37
40	.45	.98	2.43
80	.28	1.20	8.29

- If H_1 is true with a difference between β_1 and β_2 of about .3, sample sizes of 40 are sufficient to obtain a BF_{10} that is larger than 1.0 and substantially different from the corresponding Bayes factor for populations where H_0 is true.

For the comparison of H_1 with H_{1_c}, the situation in which $\beta_1 = \beta_2$ is a point on the boundary between H_1 and H_{1_c}, that is, the combination of regression coefficients closest to H_1 that no longer belongs to H_1. Stated otherwise, H_1 does not hold if $\beta_1 = \beta_2$. The following features can be observed in Table 5.20:

- For boundary points it holds that BF_{11_c} equals 1.0, that is, there is no preference for either hypothesis.

- A difference between the regression coefficients of about .2 can reliably be detected using sample sizes of 20 and larger. Note that *reliably* means that the Bayes factors are larger than 1.0 and substantially different from the 1.0 observed for the least favorable points.

- A difference between the regression coefficients of about .3 can reliably be detected using sample sizes of 12 and larger.

5.9.3   Hypotheses Evaluation Using BIEMS

The data in Table 5.12 contain the responses of 20 persons to *Income*, *IQ*, and *SES*. This is not enough for a reliable comparison of H_1 with H_0 for the effect sizes displayed in Table 5.19. BIEMS rendered $BF_{10} = 1.50/1.21 = 1.24$. This is both in terms of unconditional (it is well within the range of Bayes factors displayed for $N = 20$ in Table 5.19) as in terms of conditional error probabilities an unremarkable number; that is, it is not at all clear whether or not H_1 should be preferred over H_0.

A sample size of 20 is sufficient for a reliable comparison of H_1 with H_{1_c}. BIEMS rendered $BF_{11_c} = 2.95$. The corresponding $PMPs$ for H_1 and H_{1_c} are .74 and .26, respectively. This

TABLE 5.20: Sample Size Determination for the Comparison of Regression Coefficients (Table Entries are BF_{11_c})

	$R^2 = .32$ and $\rho_{IQ,SES} = 0$		
Sample Size	$\beta_1 = .4, \beta_2 = .4$	$\beta_1 = .48, \beta_2 = .3$	$\beta_1 = .53, \beta_2 = .2$
12	1	1.92	3.06
20	1	2.52	5.63
40	1	4.41	17.34
80	1	9.51	102.01
	$R^2 = .50$ and $\rho_{IQ,SES} = 0$		
Sample Size	$\beta_1 = .5, \beta_2 = .5$	$\beta_1 = .58, \beta_2 = .4$	$\beta_1 = .64, \beta_2 = .3$
12	1	1.99	3.76
20	1	2.92	8.22
40	1	5.46	34.82
80	1	14.72	459.05
	$R^2 = .50$ and $\rho_{IQ,SES} = .5$		
Sample Size	$\beta_1 = .41, \beta_2 = .41$	$\beta_1 = .51, \beta_2 = .3$	$\beta_1 = .59, \beta_2 = .2$
12	1	1.58	2.45
20	1	2.16	4.21
40	1	3.68	13.40
80	1	7.77	86.56

constitutes some, but not yet convincing evidence in favor of H_1 over H_{1_c}. All in all, it is not at all clear whether or not IQ is a stronger predictor of *Income* than *SES*.

Table 5.21 presents the BIEMS command files that are needed to compute BF_{1a} and BF_{11_c} (top panel) and BF_{0a} (bottom panel), which may subsequently be combined to render $BF_{10} = BF_{1a}/BF_{0a}$. Note that: #inequal=1 in the top panel denotes that H_1 consists of 1 inequality constraint; that #equal=1 in the bottom panel denotes that H_0 consists of 1 equality constraint; that the string 0 1 -1 followed by 0 in the top panel states that $0 \times \beta_0 + 1 \times \beta_1 - 1 \times \beta_2 > 0$, which is equivalent to the first component of H_1; and that the same string in the bottom panel states that $0 \times \beta_0 + 1 \times \beta_1 - 1 \times \beta_2 = 0$, which is equivalent to the first component of H_0.

5.10   Discussion

5.10.1   A Drawback

In this chapter it was illustrated for a number of specific models, what sample sizes are needed in order to be able to reliably evaluate the informative hypotheses of interest. The approach used to determine these sample sizes used data that correspond perfectly to the population from which they are sampled, that is, sample means, residual variances, and regression coefficients are equal to their population counterparts. Although illustrative, and fast to compute, there is one drawback to this approach: sample fluctuations are ignored. It is rather unusual for a data set to perfectly correspond to a population. Consequently, the sample sizes displayed in the tables in this chapter constitute only a rough but nevertheless useful indication of the sample sizes needed to evaluate the hypotheses under consideration.

In the next chapter, actual error probabilities will be computed, that is, sample fluctuations are explicitly accounted for. Although theoretically that approach is preferred, it is

TABLE 5.21: BIEMS Command Files for Hypotheses Comparing Regression Coefficients in the Context of Multiple Regression

Computation of BF_{1a} and BF_{11_c}					
input_BIEMS.txt					
Input 1:	#DV	#cov	#Tcov	N	iseed
	1	2	0	20	1111
Input 2:	#inequal	#equal	#restr		
	1	0	-1		
Input 3:	sample size	maxBF steps	scale		
	-1	-1	-1		
Input 4:	Z(DV)	Z(IV)			
	1	1			
Input 5:	Z(IV)				

inequality_constraints.txt	
R_m	r_m
0 1 -1	0

Computation of BF_{0a}					
input_BIEMS.txt					
Input 1:	#DV	#cov	#Tcov	N	iseed
	1	2	0	20	1111
Input 2:	#inequal	#equal	#restr		
	0	1	-1		
Input 3:	sample size	maxBF steps	scale		
	-1	-1	-1		
Input 4:	Z(DV)	Z(IV)			
	1	1			
Input 5:	Z(IV)				

equality_constraints.txt	
S_m	s_m
0 1 -1	0

currently so time consuming that in many situations the approach presented in this chapter, or common sense inspired by the results presented in this and the next chapter, are the only resorts available for researchers who want to determine the sample sizes needed in order to be able to compute reliable Bayes factors.

5.10.2 Informative Hypothesis versus Null or Complement?

An important feature that has not yet explicitly been discussed, can be observed throughout this chapter: the sample sizes needed for a reliable comparison of an informative hypothesis with its complement are smaller than the sample sizes needed for a reliable comparison of an informative hypothesis with the null hypothesis. This may motivate researchers to step away from the null hypothesis if it does not correspond to a population that is a plausible option as the generator of the data at hand, and to evaluate "whether their expectation is supported by the data" or "not," that is, evaluate the informative hypothesis by comparing it to its complement.

An important course of action is open to further increase the reliability with which the informative hypothesis is evaluated: further specify the complementary hypothesis. Consider, for example, the informative hypothesis $H_1 : \mu_1 < \mu_2 < \mu_3 < \mu_4$. This hypothesis represents one of the 24 ways in which 4 means can be ordered. Stated otherwise, the com-

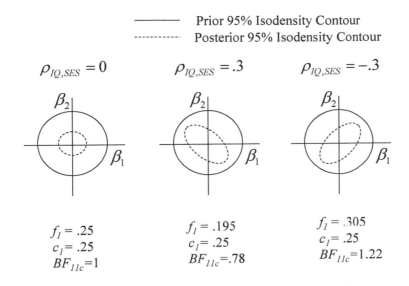

FIGURE 5.2: BF_{11_c} for regression coefficients on the boundary of H_1 and H_{1_c}

plementary hypothesis H_{1_c} : not H_1 represents the 23 remaining ways in which 4 means can be ordered. However, it will often be obvious that many of these 23 orderings do not represent realistic descriptions of the population of interest. It may, for example, be that that only $H_2 : \mu_2 < \mu_1 < \mu_4 < \mu_3$ and $H_3 : \mu_3 < \mu_1 < \mu_2 < \mu_4$ are serious competitors of H_1. In such a situation a researcher is better of evaluating a set of three competing hypotheses consisting of H_1, H_2 and H_3. The comparison of a hypothesis with its complement is less reliable than the comparison of competing hypotheses, because the latter are more specific and thus more prone to falsification than the first. This will further be illustrated in the next chapter where the evaluation of competing hypotheses will be addressed in the context of repeated measures analysis.

5.11 Appendix 5A: Bayes Factors for Parameters on the Boundary of H_1 and H_{1_c}

In Figure 5.2 the left-hand and middle figures correspond to the first two panels in Table 5.15, the right-hand figure displays a situation in which $\rho_{IQ,SES} = -.3$. As can be seen in the left hand figure, the proportion of the prior distribution in agreement with H_1 is .25. If $\beta_1 = \beta_2 = 0$, the proportion of the resulting posterior distribution in agreement with H_1 is also .25. The result is $BF_{11_c} = 1$. In this situation, Bayes factors larger than 1 indicate support for H_1 and values smaller than 1 support for H_{1_c}.

However, this intuitively appealing feature only arises if the posterior correlation between the parameters that are constrained is 0 (like in the left-hand figure in Figure 5.2, and, in the ANOVA and ANCOVA examples presented and discussed earlier in this chapter). As shown in the middle and right-hand figures, the Bayes factor for a boundary value is .78

if the correlation is .3 and 1.22 if the correlation is $-.3$. Stated otherwise, the value BF_{11_c} obtained for regression coefficients on the boundary of H_1 and H_{1_c} depends on the posterior correlation of the parameters involved in the constraints. As can be seen in the middle figure, if the correlation between IQ and SES is .3, the posterior correlation between β_1 and β_2 is negative. The consequence is that "only" 19.5% of the posterior distribution is in agreement with H_1, that is, the Bayes factor is .78. Stated otherwise, $\beta_1 = \beta_2 = 0$ combined with a positive correlation between IQ and SES constitutes evidence against H_1. Using the right-hand figure, it can similarly be explained that $\beta_1 = \beta_2 = 0$ combined with a negative correlation between IQ and SES constitutes evidence in favor of H_1. In both situations, it is still the case that Bayes factors larger than 1 indicate support for H_1 and values smaller than 1 indicate support for H_{1_c}. However, due to the posterior correlation, a Bayes factor of 1 is no longer obtained on the natural boundary between H_1 and H_{1_c}.

5.12   Appendix 5B: Command Files for GenMVLData

Table 5.22 containts the command file input.txt needed to run GenMVLData such that it renders the data matrix suited for ANOVA that was analyzed in Table 5.4 for $N = 20$ and $e = .2$. Command files for GenMVLData may be constructed manually or using a Windows user interface. Further information with respect to GenMVLData can be found in Section 9.2.4.

The first two lines in input.txt state: one dependent variable #DV=1, four groups #groups=4, no covariates or predictors #cov=0, and that the data should have sample means and within group variance equal to the corresponding population quantities exact=1. With exact data, the command iseed=-1 is inactive.

The second group of two lines list that the sample size in each group is equal to 20. The third group lists that the within group *standard deviation* is equal to 1 (do not by mistake enter the within group variance; what is needed here is the square root of the within group variance). The fourth group is inactive if #DV=1. The group of five lines lists that the population means in the four groups are 0 .2 .4 .6.

Table 5.23 contains the command file input.txt needed to run GenMVLData such that it renders the data matrix suited for ANCOVA that was analyzed in Table 5.7 for $N = 20$ and $d = .2$. The main changes compared to the command file for ANOVA displayed in Table 5.22 are: #cov=1 to generate scores on a covariate; an extra line in the parameter matrix .6 that states that within each group the regression coefficient relating the covariate to the dependent variable is equal to .6; the columns of four 0s and 1s denote that within each of the four groups the covariate is normally distributed with a mean of 0 and a variance of 1; and the residual variance is .64, that is, the standard deviation of the error terms is $\sqrt{.64} = .80$. Note that a standardized covariate with a regression coefficient of .6 contributes $.6^2 = .36$ to the variance of the dependent variable. If in addition the *residual variance* is set at .64, the variance of the dependent variable will turn out to be $.36 + .64 = 1$ in each group.

Table 5.24 contains the command file input.txt needed to run GenMVLData such that it renders the data matrix suited for multiple regression that was analyzed in Table 5.11 for $N = 20$, $\rho_{IQ,SES} = .3$, and $\beta_0 = 0, \beta_1 = .20, \beta_2 = .20$. In addition to the discussion of input.txt for ANOVA and ANCOVA, only a few comments are necessary. There is now one group #groups=1 and two predictors #cov=2. Both predictors have a bivariate normal distribution with zero means, unit variances, and a covariance of .3. Standardized covariates that have a mutual correlation of .3 and a regression coefficient of .20 contribute .10 to the variance of the dependent variable. If in addition the *residual variance* is set at .90 (that is, the standard deviation is .948), the variance of the dependent variable will turn out to be .10 + .90 = 1.

TABLE 5.22: input.txt Generating the ANOVA Data for $N = 20$ and $e = .2$ in Table 5.4

```
#DV #groups #cov iseed exact
1    4      0    -1    1

group sizes
20 20 20 20

standard deviations for the error terms
1

correlation matrix for the error terms
1

parameter matrix
0
 .2
 .4
 .6

means for normal covariates per group

covariance matrices for normal covariates per group
```

TABLE 5.23: input.txt Generating the ANCOVA Data for $N = 20$ and $e = .2$ in Table 5.7

```
#DV #groups #cov iseed exact
1    4       1     -1    1

group sizes
20 20 20 20

standard deviations for the error terms
.80

correlation matrix for the error terms
1

parameter matrix
0
.2
.4
.6
.6

means for normal covariates per group
0
0
0
0

covariance matrices for normal covariates per group
1
1
1
1
```

TABLE 5.24: input.txt Generating the Multiple Regression Data for $N = 20$, $\rho_{IQ,SES} = .3$ and $\beta_0 = 0, \beta_1 = .20, \beta_2 = .20$ in Table 5.11

```
#DV #groups #cov iseed exact
1    1       2    -1     1

group sizes
20

standard deviations for the error terms
.948

correlation matrix for the error terms
1

parameter matrix
0
.2
.2

means for normal covariates per group
0 0

covariance matrices for normal covariates per group
 1 .3
.3  1
```

Chapter 6

Sample Size Determination: The Multivariate Normal Linear Model

6.1 Introduction

The discussion of sample size determination that started in Chapter 5 continues in this chapter. Two features will be changed. First of all, the discussion continues using the examples of informative hypotheses that were introduced in Chapter 2 in the context of the multivariate normal linear model. Using these examples, three situations are distinguished: the evaluation of one informative hypothesis in Sections 6.3 and 6.4, the evaluation of a series of informative components in Section 6.5, and the evaluation of competing hypotheses in Section 6.6. Each application contains the same steps as in Chapter 5:

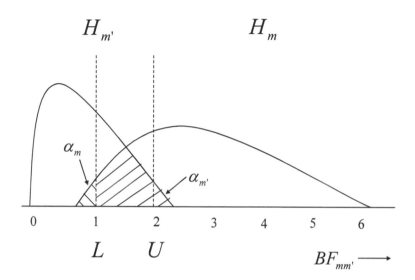

FIGURE 6.1: Distribution of the Bayes factor under H_m and $H_{m'}$.

- Step 1: Formulate the informative hypotheses.

- Step 2: Sample size determination.

- Step 3: Evaluate the Bayes factors and posterior model probabilities for the data at hand.

Second, sample size determination will no longer be based on Bayes factors computed for data matrices that correspond perfectly to the generating population. In this chapter actual error probabilities are computed using multiple data matrices sampled from the populations of interest. These error probabilities are introduced in the next section.

6.2   Sample Size Determination: Error Probabilities

In Figure 6.1, the label H_m represents a population where H_m is true. Note that the label m may refer to an informative hypothesis, its complement, a null hypothesis, or an unconstrained hypothesis. The label $H_{m'}$ represents a population where $H_{m'}$ is true. From both populations a sequence of data matrices can be sampled. For each of these data matrices $BF_{mm'}$ can be computed, its distribution under H_m and $H_{m'}$ is displayed in the figure. Two error probabilities are indicated: the proportion of incorrect conclusions if H_m is true is about 2.5%, that is, in 2.5% of the cases $BF_{mm'}$ is smaller than 1 (the shaded area below 1); and the proportion of incorrect conclusions if $H_{m'}$ is true is about 20%, that is, in 20% of the cases $BF_{mm'}$ is larger than 1 (the shaded area above 1). Note that use of the reference value $BF_{mm'} = 1$ is in line with Principle 5.1 presented in Chapter 5.

Principle 5.2 from Chapter 5 stated that reliable Bayes factors are only obtained if the values obtained under H_m and $H_{m'}$ are substantially different. This principle is also relevant if the reliability of the Bayes factor is determined by means of error probabilities. Suppose, for example, that a researcher wants to evaluate $H_m : \mu_1 < \mu_2 < \mu_3 < \mu_4$ versus $H_{m'}$: not H_m. Suppose also that based on a sample of 15 per group, the result is a Bayes factor of $BF_{mm'} = 5$ in favor of H_m. The straightforward (and correct) interpretation is that the support in the data for H_m is 5 times larger than the support for $H_{m'}$. However, when generalizing this result to the population of interest, the reliability of the Bayes factor must be considered. To do this, the following questions must be answered: what values of the Bayes factor will be observed if data sets of the current size are sampled from populations where H_m holds?; and the same question but then with respect to $H_{m'}$. Hypothetical results for data matrices repeated sampled from populations where H_m and $H_{m'}$ hold, are displayed in Figure 6.1. As can be seen, values of $BF_{mm'}$ smaller than 1 are clearly indicative of $H_{m'}$ because the proportion of $BF_{mm'}$ smaller than 1 if H_m is true is rather small (about 2.5%); values larger than 2 are clearly indicative of H_m because the proportion of $BF_{mm'}$ larger than 2 if $H_{m'}$ is true is rather small (about 2%); and values of $BF_{mm'}$ between 1 and 2 could result from both H_m and $H_{m'}$, that is, these values cannot reliably be used to distinguish between H_m and $H_{m'}$. The range from lower bound $L = 1$ to upper bound $U = 2$ in Figure 6.1 will be called the interval of indecision. The proportion of errors if H_m is true will be denoted by α_m, that is, the proportion of $BF_{mm'}$ smaller than L if H_m is true. The proportion of errors if $H_{m'}$ is true will be denoted by $\alpha_{m'}$, that is, the proportion of $BF_{mm'}$ larger than U if $H_{m'}$ is true. Note that α_m and $\alpha_{m'}$ are unconditional error probabilities. For the example at hand, the case is clear-cut: $BF_{mm'} = 5$. This is larger than 2, that is, it can reliably be concluded that the support in the data is larger for H_m than for $H_{m'}$. However, if a value of $BF_{mm'}$ between 1 and 2 had been obtained, a reliable evaluation of the hypotheses of interest would not have been possible.

As exemplified in the previous paragraph, the result of sample size determination based on error probabilities is of the following form:

- If $BF_{mm'} < L$, it can reliably be concluded that $H_{m'}$ is preferred.

- If $L < BF_{mm'} < U$, there is no reliable preference for either H_m or $H_{m'}$.

- IF $BF_{mm'} > U$, it can reliably be concluded that H_m is preferred.

Note that the lower bound L and the upper bound U should be chosen such that the error probabilities with respect to these bounds are acceptable. In line with Principle 5.3 presented in Chapter 5, new and arbitrary guidelines like ".05" are avoided if each researcher has to determine for himself which error probabilities are acceptable. In the sections that follow, the author of this book chooses L and U such that $\alpha_m = \alpha_{m'} = .10$. But other researchers may very well prefer other values. Note also that for $N \to \infty$, both the lower and the upper bound will converge to 1, that is, for large sample sizes, there is no interval of indecision.

6.3   Multivariate One-Sided Testing

6.3.1   Hypotheses

In Section 2.3, a hypothetical researcher was interested in the improvement in depression, anxiety, and social phobia of 20 persons after receiving therapy for their complaints. Note that a score of 1 on one of the outcome measures implies that, for example, depression was reduced by 1 unit (each of the three variables was measured on a scale running from 1 to 20 before and after receiving therapy for 12 weeks), and that the score -2 implies that, for example, depression increased by 2 units. The researcher wanted to know whether therapy leads to a decrease in depression, anxiety, and social phobia or not. Stated otherwise, his hypothesis of interest was

$$H_1 : \quad \begin{matrix} \mu_1 > 0 \\ \mu_2 > 0 \\ \mu_3 > 0 \end{matrix} \quad . \tag{6.1}$$

This hypothesis can be compared to the corresponding null hypothesis, that is,

$$H_0 : \quad \begin{matrix} \mu_1 = 0 \\ \mu_2 = 0 \\ \mu_3 = 0 \end{matrix} \quad , \tag{6.2}$$

or to its complement, that is, H_{1_c} : not H_1. Note that μ_1, μ_2, and μ_3 denote the mean improvement in depression, anxiety, and social phobia, respectively.

6.3.2   Sample Size Determination

When determining sample sizes, there are always many variations in parameter values, that is, populations of interest, that can be considered. Therefore, such that the overview is not lost, the variation in parameter values and populations will be kept to a minimum. Here, sample sizes of 25 and 50 will be considered. Furthermore, effect sizes of 0, .3, and .5 will be considered, and, correlations ρ between the three dependent variables of .3 and .6. This resulting simulation design contains $2 \times 3 \times 2 = 12$ cells (see Tables 6.1 and 6.2). The effect size that will be used is Cohen's d, that is, $e_p = \mu_p/\sigma_p$ for $p = 1, \ldots, 3$, where μ_p and σ_p denote the population mean and standard deviation of the p-th dependent variable, respectively. Note that e is used to denote Cohen's d if it applies to sample or population effect sizes, and that d is used to denote Cohen's d if it is used to formulate an informative hypothesis using effect sizes.

In each cell of the design, 1,000 data matrices are sampled from the corresponding population. In Appendix 6A the GenMVLData command file used is presented for $\rho = .3$, $N = 25$, and $e = .3$. Note the use of the command exact=0, which denotes that a data matrix has to be sampled from the population of interest, that is, the sample means and sample variances are *not* exactly the same as in the population. After sampling 1,000 data matrices containing 25 and 50 persons from each of the six populations under consideration, BIEMS can be used to compute BF_{10} and BF_{11_c} for each data matrix. A summary of the results is displayed in Tables 6.1 and 6.2.

The left-hand panel of Table 6.1 shows the error probabilities α_0 (incorrectly rejecting H_0 if it is true) and α_1 (incorrectly rejecting H_1 if it is true) if H_0 is chosen if $BF_{10} < 1$, and H_1 is chosen if $BF_{10} > 1$. As can be seen, for H_0 and H_1 with $e = .5$, the error probabilities are about 10% or smaller. These are acceptable numbers, which implies that comparison of

TABLE 6.1: Reliability of BF_{10} for Multivariate-One Sided Testing

		$\rho = .3$					
Sample Size	$e = 0$	$e = .3$	$e = .5$	$e = 0$	$e = .3$	$e = .5$	
25	10% > 1	30% <1	3% < 1		10%< .31		
50	5% > 1	17% <1	0%<1		10%< .49		
		$\rho = .6$					
Sample Size	$e = 0$	$e = .3$	$e = .5$	$e = 0$	$e = .3$	$e = .5$	
25	8%>1	40% <1	11%<1		10%< .22		
50	5%>1	31% <1	3%<1		10%< .23		

BF_{10} with a threshold value of 1.0 renders a procedure with acceptable unconditional error probabilities.

However, looking at H_1 with $e = .3$, it can be seen that the error probabilities are rather large. Even if H_1 is true, in up to 40% of the cases it is incorrectly concluded that H_0 is the best hypothesis. The right-hand panel of Table 6.1 displays the threshold values for BF_{10} that would render error probabilities of 10% if $e = .3$. These imply that the decision to reject H_1 in favor of H_0 should not be taken if BF_{10} is smaller than 1 (then the error probability is too large) but if BF_{10} is smaller than, say, .30.

Within the context of sample sizes and populations displayed in Table 6.1, the following decision rules seem reasonable:

- Prefer H_0 over H_1 if $BF_{10} < .30$.

- Accept that the data are not decisive if $.30 < BF_{10} < 1.0$.

- Choose H_1 over H_0 if $BF_{10} > 1.0$.

Note that sample size determination, like classical power analysis, is not an exact science but a serious attempt to gain information with respect to the sample sizes needed in order to obtain reliable Bayes factors (in terms of error probabilities) for the empirical data that will be collected. The threshold of .30 appearing in the decision rule is not a golden standard, but "a number that is in the ballpark" if

- Error probabilities of 10% are considered acceptable.

- The smallest effect size of interest is .3. Note that if the smallest effect size of interest would have been .5, the number .30 would be replaced by 1, that is, there would not have been an area of indecision.

- The sample sizes of interest are between 25 and 50.

- The correlations between the dependent variables are between .3 and .6.

Note furthermore that 10% is also *not* a golden standard. It is the personal opinion of the author of this book that error probabilities of 10% under each of the hypotheses of interest are acceptable. However, others may very well opt for error probabilities of 6% or any other percentage, as long as 5% is avoided.

Results for the comparison of H_1 with H_{1_c} are displayed in Table 6.2. Displayed are two populations in agreement with H_1 with, respectively, $e = .3$ and $e = .5$. The population with $e = 0$ is on the boundary between H_1 and H_{1_c}. This implies that Table 6.2 displays a worst-case scenario, that is, how well can H_1 be distinguished from H_{1_c} if populations with different effect sizes from H_1 are compared to the population from H_{1_c} that is closest to H_1.

TABLE 6.2: Reliability of BF_{11_c} for Multivariate One-Sided Testing

| | | | | $\rho_{y_p,y_{p'}} = .3$ | | |
Sample Size	$e = 0$	$e = .3$	$e = .5$	$e = 0$	$e = .3$	$e = .5$
25	50% >1	3% <1	0% <1	10%>5.11		
50	52% >1	1% <1	0% <1	10%>6.44		

| | | | | $\rho_{y_p,y_{p'}} = .6$ | | |
Sample Size	$e = 0$	$e = .3$	$e = .5$	$e = 0$	$e = .3$	$e = .5$
25	61% >1	2% <1	0% <1	10%> 6.89		
50	64% >1	0%<1	0% <1	10%> 8.69		

As can be seen in Table 6.2 if BF_{11_c} is compared to a threshold value of 1, the error probabilities are very small if H_1 is true. However, the error probabilities for the population on the boundary between H_1 and H_{1_c} can attain values up to 60%. This implies that a decision rule based on a threshold value of 1 does not render reliable Bayes factors, that is, a reliable decision with respect to the support in the data for H_1 and H_{1_c}. However, if error probabilities of about 10% are considered acceptable, the following decision rules can be used:

- Choose H_{1_c} if $BF_{11_c} < 1.0$.

- Accept that the data are not decisive if $1.0 < BF_{10} < 7.0$.

- Choose H_1 if $BF_{11_c} > 7.0$.

Note that these rules only apply to the populations considered in Table 6.2. This implies that the interval of indecision will be much smaller if populations in the interior of H_{1_c} instead of on the boundary between H_1 and H_{1_c} are considered.

TABLE 6.3: Example Data for the Therapy Group

Depression	Anxiety	Social Phobia	Group
−1	−2	−1	1
1	1	1	1
0	2	0	1
2	1	1	1
3	2	1	1
−1	−1	1	1
0	1	2	1
0	0	−1	1
2	1	2	1
1	0	0	1
−1	−1	1	1
1	1	1	1
0	1	0	1
2	1	2	1
3	2	1	1
−1	−1	0	1
0	1	2	1
0	0	−1	1
2	1	2	1
1	−1	1	1

TABLE 6.4: Sample Means and Standard Errors for the Therapy and Control Group

Group		Depression	Anxiety	Social Phobia
Therapy	Mean	.70	.45	.75
	SE	.29	.26	.23
Control	Mean	−.30	−.25	−.25
	SE	.18	.16	.12

6.3.3 Hypotheses Evaluation Using BIEMS

After an inspection of the sample size determination in the previous section, it was decided to use a sample of 20 persons in the therapy group. The data in Table 6.3 were previously presented in Section 2.3. For analysis with BIEMS, the data must be placed in a text file named `data.txt`. The first three columns contain the improvement with respect to depression, anxiety, and social phobia, respectively. The last column states that all persons belong to the same and only group in the data file.

The BIEMS command files presented in Appendix 6A rendered $BF_{10} = 6.70/.69 = 9.71$ and $BF_{11_c} = 35.90$. According to the decision rules given above, both numbers can be reliably used to conclude that H_1 is strongly supported by the data. The corresponding conditional error probabilities are $PMPs$ of .90 and .10 for H_1 versus H_0, respectively, and .97 and .03 for H_1 versus H_{1_c}. Both represent substantial support for H_1. Note that these results are in line with the sample means for the variables involved as displayed in Table 6.4. It can be concluded that therapy has a beneficial effect on depression, anxiety, and social phobia.

6.4 Multivariate Treatment Evaluation

6.4.1 Hypotheses

In the previous section it was investigated whether therapy had a beneficial effect on depression, anxiety, and social phobia. In this section it will be investigated whether the decrease in depression, anxiety, and social phobia is larger in the therapy group compared to a control group consisting of persons who were on a waiting list during the course of the therapy in the other group. The main hypothesis is

$$H_1: \begin{array}{c} \mu_{11} > \mu_{12} \\ \mu_{21} > \mu_{22} \\ \mu_{31} > \mu_{32} \end{array}, \tag{6.3}$$

that is, the means in the therapy group are larger than the corresponding means in the control group. Note that μ_{11} denotes the decrease in depression (first index equal to 1 denotes depression, 2 and 3 denote anxiety and social phobia, respectively) in the therapy group (second index equal to 1 denotes therapy group, 2 denotes control group).

A traditional evaluation of this informative hypothesis is obtained if it is compared to the null hypothesis:

$$H_0: \begin{array}{c} \mu_{11} = \mu_{12} \\ \mu_{21} = \mu_{22} \\ \mu_{31} = \mu_{32} \end{array}. \tag{6.4}$$

In line with the ideas presented in this book, is a comparison of if H_1 with H_{1_c}: not H_1.

TABLE 6.5: Reliability of BF_{10} for Multivariate Treatment Evaluation

			$\rho = .3$			
Sample Size	$e = 0$	$e = .3$	$e = .5$	$e = 0$	$e = .3$	$e = .5$
25	15% >1	39%<1	12%<1	10% > 1.52	10% < .29	
50	10% >1	31%<1	4%<1		10% <.29	
			$\rho = .6$			
Sample Size	$e = 0$	$e = .3$	$e = .5$	$e = 0$	$e = .3$	$e = .5$
25	12% >1	50%<1	21%<1	10% > 1.15	10% <.25	10% <.58
50	8% >1	40%<1	10%<1		10% <.23	

6.4.2   Sample Size Determination

The sample sizes, effect sizes, and correlations among the three dependent variables under consideration must be specified because they all influence the reliability of the Bayes factor. The effect size that is used here is Cohen's d, that is, $e_p = (\mu_{p1} - \mu_{p2})/\sigma_p$ for $p = 1, \ldots, 3$, where μ_{pj} denotes the mean of the j-th group on the p-th dependent variable and σ_p the within group variance of the p-th dependent variable. When determining sample sizes, there are always many variations in parameter values, that is, populations of interest, that can be considered. To keep the overview, the variation in parameter values and populations will be kept to a minimum. Here, sample sizes of 25 and 50 are considered. Furthermore, effect sizes of 0, .3, and .5 are considered, and correlations between the three dependent variables of $\rho = .3$ and $\rho = .6$.

In each cell of the design 1,000 data matrices are sampled from the corresponding population. In Appendix 6B the `GenMVLData` command file used is presented for $\rho = .3$, $N = 25$ per group, and $e = .3$. Note the use of the command `exact=0`, which denotes that a data matrix must be sampled from the population of interest, that is, the sample means and sample variances are *not* exactly the same as in the population. After sampling 1,000 data matrices containing $2 \times 25 = 50$ and $2 \times 50 = 100$ persons from each of the six populations under consideration, `BIEMS` can be used to compute BF_{10} and BF_{11_c} for each data matrix. A summary of the results is displayed in Tables 6.5 and 6.6.

The left-hand panel of Table 6.5 shows the error probabilities α_0 (incorrectly rejecting H_0 if it is true) and α_1 (incorrectly rejecting H_1 if it is true) if H_0 is chosen if $BF_{10} < 1$, and H_1 is chosen if $BF_{10} > 1$. The right-hand panel of Table 6.5 displays the threshold values for BF_{10} that would render error probabilities of 10% for the corresponding cells in the left-hand panel that have unsatisfactory error probabilities.

For sample sizes of 25 per group, the following decision rules seem appropriate:

- Prefer H_0 over H_1 if $BF_{10} < .25$.

- Accept that the data are not decisive if $.25 < BF_{10} < 1.35$.

- Choose H_1 over H_0 if $BF_{10} > 1.35$.

For sample sizes of 50 per group, the decision rules are

- Prefer H_0 over H_1 if $BF_{10} < .25$.

- Accept that the data are not decisive if $.25 < BF_{10} < 1.00$.

- Choose H_1 over H_0 if $BF_{10} > 1.00$.

Results for the comparison of H_1 with H_{1_c} are displayed in Table 6.6. Displayed are two populations in agreement with H_1 with, respectively, $e = .3$ and $e = .5$. Note again that

TABLE 6.6: Reliability of BF_{11_c} for Multivariate Treatment Evaluation

			$\rho = .3$			
Sample Size	$e = 0$	$e = .3$	$e = .5$	$e = 0$	$e = .3$	$e = .5$
25	$50\% > 1$	$9\% < 1$	$1\% < 1$	$10\% > 4.15$		
50	$50\% > 1$	$4\% < 1$	$0\% < 1$	$10\% > 5.16$		

			$\rho = .6$			
Sample Size	$e = 0$	$e = .3$	$e = .5$	$e = 0$	$e = .3$	$e = .5$
25	$60\% > 1$	$7\% < 1$	$1\% < 1$	$10\% > 5.50$		
50	$63\% > 1$	$1\% < 1$	$0\% < 1$	$10\% > 7.87$		

the population with $e = 0$ is on the boundary between H_1 and H_{1_c}. This implies that Table 6.6 displays a worst-case scenario, that is, how well can H_1 be distinguished from H_{1_c} if populations with different effect sizes from H_1 are compared to the population from H_{1_c} that is closest to H_1.

As can be seen in Table 6.6, if BF_{11_c} is compared to a threshold value of 1, the error probabilities are very small if H_1 is true. However, the error probabilities for the population on the boundary between H_1 and H_{1_c} can attain values up to 60%. This implies that a decision rule based on a threshold value of 1 does not render reliable Bayes factors, that is, a reliable decision with respect to the support in the data for H_1 and H_{1_c}. However, if error probabilities of about 10% are considered acceptable, the following decision rules can be used:

- Choose H_{1_c} if $BF_{11_c} < 1.0$.

- Accept that the data are not decisive if $1.0 < BF_{10} < 5.5$.

- Choose H_1 if $BF_{11_c} > 5.5$.

Note that these rules only apply to the populations considered in Table 6.6. This implies that the interval of indecision will be much smaller if populations in the interior of H_{1_c} instead of on the boundary between H_1 and H_{1_c} are considered.

6.4.3 Hypothesis Evaluation Using BIEMS

After an inspection of the sample size determination in the previous section, it was decided to use a sample of 20 persons in both the therapy and the control group. The data in Tables 6.3 and 6.7 were previously presented in Sections 2.3 and 2.4, respectively. For analysis with BIEMS, these data must be placed in one text file named data.txt. The first three columns contain the scores on depression, anxiety, and social phobia, respectively. The last column states that persons in therapy belong to group 1 and that the control persons belong to group 2.

The BIEMS command files presented in Appendix 6B rendered $BF_{10} = BF_{1a}/BF_{0a} = 6.82/.14 = 48.71$ and $BF_{11_c} = 39.06$. According to the decision rules given above, both numbers can be reliably used to conclude that H_1 is strongly supported by the data. The corresponding conditional error probabilities are PMPs of .98 and .02 for H_1 versus H_0, respectively, and .97 and .03 for H_1 versus H_{1_c}. Both represent substantial support for H_1, that is, the decrease in depression, anxiety, and social phobia is larger in the therapy group than in the control group. Note that this is in line with the sample means for the variables involved as displayed in Table 6.4.

TABLE 6.7: Example Data for the Control Group

Depression	Anxiety	Social Phobia	Group
0	0	−1	2
−1	1	0	2
−1	−1	0	2
0	0	0	2
1	−1	0	2
−2	0	−1	2
1	1	0	2
0	−1	−1	2
0	1	0	2
0	−1	0	2
−1	−1	−1	2
−1	0	−1	2
1	0	0	2
0	0	0	2
−1	0	0	2
−1	0	0	2
0	−1	0	2
0	0	1	2
0	−1	−1	2
−1	−1	0	2

6.5   Multivariate Regression

6.5.1   Hypotheses

In Section 2.5, a series of informative components was presented in the context of multivariate regression. Arithmetic Ability (A, in line with the tradition in the Netherlands scored on a scale from 1 = very poor to 10 = outstanding) and Language Skills (L with the same scoring as A) were predicted by the scores of the fathers (FA and FL) and mothers (MA and ML) of the 100 children on A and L, respectively. It is important to note that the two dependent variables and the four predictors are standardized and, consequently, that the βs appearing in (6.5) are standardized. The equations corresponding to this design are

$$Z(A_i) = \beta_{A0} + \beta_{A1}Z(FA_i) + \beta_{A2}Z(FL_i) + \beta_{A3}Z(MA_i) + \beta_{A4}Z(ML_i) + \epsilon_{Ai}$$

(6.5)

$$Z(L_i) = \beta_{L0} + \beta_{L1}Z(FA_i) + \beta_{L2}Z(FL_i) + \beta_{L3}Z(MA_i) + \beta_{L4}Z(ML_i) + \epsilon_{Li}.$$

The first hypothesis is the reference hypothesis:

$$H_1: \begin{array}{l} \beta_{A1} > 0 \\ \beta_{A3} > 0 \\ \beta_{L2} > 0 \\ \beta_{L4} > 0 \end{array}.$$

(6.6)

There is a series of three informative components H_2, H_3, and H_4 that are potential extensions of H_1:

$$H_{1\&2}: \begin{matrix} H_1 \\ \beta_{A1} > \beta_{A2} \\ \beta_{A3} > \beta_{A4} \\ \beta_{L2} > \beta_{L1} \\ \beta_{L4} > \beta_{L3} \end{matrix}, \tag{6.7}$$

$$H_{1\&3}: \begin{matrix} H_1 \\ \beta_{A1} > \beta_{L1} \\ \beta_{A3} > \beta_{L3} \\ \beta_{L2} > \beta_{A2} \\ \beta_{L4} > \beta_{A4} \end{matrix}, \tag{6.8}$$

and

$$H_{1\&4}: \begin{matrix} H_1 \\ \beta_{A2} > \beta_{L1} \\ \beta_{A4} > \beta_{L3} \end{matrix}. \tag{6.9}$$

TABLE 6.8: Description of Populations in Agreement with H_1 and H_1 through H_4

Population in Agreement with H_1					
	β_{p0}	β_{p1}	β_{p2}	β_{p3}	β_{p4}
$p = A$	0	.20	.20	.20	.20
$p = L$	0	.20	.20	.20	.20
Population in Agreement with H_1 through H_4					
	β_{p0}	β_{p1}	β_{p2}	β_{p3}	β_{p4}
$p = A$	0	.3	.15	.3	.15
$p = L$	0	-.17	.34	-.17	.34

Population Mean of the Four Predictors			
FA	FL	MA	ML
0	0	0	0

Covariance Matrix of the Four Predictors				
	FA	FL	MA	ML
FA	1	.3	.3	.3
FL	.3	1	.3	.3
MA	.3	.3	1	.3
ML	.3	.3	.3	1

Residual Covariance Matrices				
	H_1		H_1 through H_4	
	$p = A$	$p = L$	$p = A$	$p = L$
$p = A$	.70	.21	.60	.20
$p = L$	.21	.70	.20	.76

6.5.2 Sample Size Determination

In Table 6.8, two populations are described. The first is in agreement with H_1 and the second in agreement with H_1 through H_4. In the first population, 30% of the variance of both dependent variables is explained by the four predictors. In the second population, 40% of the variance of the first and 24% of the variance of the second dependent variable is explained by the predictors. After sampling 1,000 data matrices containing 100 persons and 1,000 data matrices containing 200 persons from both populations, various Bayes factors are computed

TABLE 6.9: Reliability of Bayes Factors for Multivariate Regression

Sample Size 100	Population H_1	Population H_1	Population H_1 through H_4	Population H_1 through H_4
$BF_{1\&2,1}$	1% > 1		15% < 1	10% < .82
$BF_{1\&3,1}$	7% > 1		3% < 1	
$BF_{1\&4,1}$	34% > 1	10% > 1.93	0% < 1	
One or More Errors	39% > 1	16% > 1, 1, 1.93	17% < 1	12% < .82, 1, 1
Sample Size 200	Population H_1	Population H_1	Population H_1 through H_4	Population H_1 through H_4
$BF_{1\&2,1}$	1% > 1		5% < 1	
$BF_{1\&3,1}$	7% > 1		0% < 1	
$BF_{1\&4,1}$	33% > 1	10% > 1.99	0% < 1	
One or More Errors	38% > 1	17% > 1, 1, 1.99	5% < 1	

for each data matrix. In Appendix 6C the `GenMVLData` command file used to simulate data in agreement with H_1 through H_4 is presented The results are summarized in Table 6.9. Sample size determination could have been elaborated via the addition of populations in agreement with $H_{1\&2}$, $H_{1\&3}$, and $H_{1\&4}$. Furthermore, for each hypothesis more than one population could have been considered. However, in order not to lose overview of what is going on, here only two populations, one representative for H_1 and one representative for $H_{1\&2\&3\&4}$, will be considered.

To determine which informative components are extensions of the reference hypothesis H_1 that are supported by the data, the following course of action is followed:

- Determine for each informative component whether or not it is a valuable extension of H_1.

- Compute the Bayes factor for the combination of all valuable informative components versus the reference hypothesis.

The top panel of Table 6.9 presents the results for a sample size of 100. For $BF_{1\&3,1}$, reliable support in the data for the extended hypothesis $H_{1\&3}$ over the reference hypothesis H_1 is found if $BF_{1\&3,1}$ is larger than 1.0. No support is found if this Bayes factor is smaller than 1.0. For $BF_{1\&2,1}$, the following decision rules are appropriate:

- Prefer $H_{1\&2}$ over H_1 if $BF_{1\&2,1} > 1$.

- Accept that the data are not decisive if $.82 < BF_{1\&4,1} < 1$.

- Choose H_1 over $H_{1\&2}$ if $BF_{1\&2,1} < .82$.

For $BF_{1\&4,1}$, the following decision rules are appropriate:

- Prefer $H_{1\&4}$ over H_1 if $BF_{1\&4,1} > 1.93$.

- Accept that the data are not decisive if $1 < BF_{1\&4,1} < 1.93$.

- Choose H_1 over $H_{1\&4}$ if $BF_{1\&4,1} < 1$.

The bottom panel of Table 6.9 presents the results for a sample size of 200. For $BF_{1\&2,1}$ and $BF_{1\&3,1}$, reliable support in the data for the extended hypotheses $H_{1\&2}$ and $H_{1\&3}$ over the reference hypothesis H_1 is found using the natural reference value 1. However, for $BF_{1\&4,1}$, the following decision rules are appropriate:

- Prefer $H_{1\&4}$ over H_1 if $BF_{1\&4,1} > 1.99$.

TABLE 6.10: Regression Coefficients and Standard Errors for the Data in Appendix 2A

Dependent Variable		β_0	β_1	β_2	β_3	β_4
A	Estimate	.00	.40	−.01	.13	−.14
	SE	.09	.10	.11	.11	.10
L	Estimate	.00	−.10	.35	−.21	.30
	SE	.09	.10	.11	.11	.10

- Accept that the data are not decisive if $1 < BF_{1\&4,1} < 1.99$.

- Choose H_1 over $H_{1\&4}$ if $BF_{1\&4,1} < 1$.

The last lines of the top and bottom panels of Table 6.9 display the error rates due to the evaluation of multiple instead of one hypothesis. As can be seen, the probability of one or more incorrect decisions is inflated, that is, larger than 10%, if the decision rules given above are used. Even if H_1 is true, another combination of hypotheses is preferred in about 17% of the cases, and if H_1 through H_4 are true, another combination of hypotheses is preferred in 12% of the cases. This phenomenon can be called "capitalization on chance," that is, the more hypotheses or components are evaluated, the higher the probability of incorrect decisions. The implication is that researchers should only include those hypotheses and components in their analyses that correspond to the expectations they have. A fishing expedition in which all kinds of hypotheses and components are evaluated is not recommended because it will substantially increase the error rates, that is, the number of erroneous decisions.

6.5.3   Hypothesis Evaluation Using BIEMS

The data for the example at hand can be found in Appendix 2A. The BIEMS command files rendering $BF_{1\&2,a}$ and $BF_{1,a}$ can be found in Appendix 6C. Note that the unconstrained prior distributions corresponding to $H_{1\&2}$ and H_1 are in their default versions not compatible. This is automatically repaired if the BIEMS Windows user interface is used. In Appendix 6C it is elaborated how this situation can be handled if the stand alone version of BIEMS is used. From both Bayes factors it is easy to obtain the Bayes factor of interest: $BF_{1\&2,1} = BF_{1\&2,a}/BF_{1,a}$. The other Bayes factors of interest are obtained analogously. For the data in Appendix 2A the following Bayes factors are obtained: $BF_{1\&2,1} = 42.32/13.72 = 3.08$, $BF_{1\&3,1} = 3.14$, and $BF_{1\&4,1} = 0$. From these results it can reliably be concluded that extensions of H_1 with H_2 and H_3 is supported by the data, but that extension with H_4 is not. The joint evidence in favor of H_1 extended with H_2 and H_3 is $BF_{1\&2\&3,1} = 4.58$. Note that these results are in line with the regression coefficients obtained using SPSS (http://www.spss.com) as presented in Table 6.10. Note that $PMP_{1\&2\&3} = .82$ and that $PMP_1 = .18$, that is, H_1 cannot yet be discarded completely, especially because the error rates are increased due to capitalization on chance. Nevertheless, there is some evidence that H_2, which states that the arithmetic skills of the parents have a stronger relation to the arithmetic skills of the child than the language skills of the parent, and H_3 which states that arithmetic ability of the parents is a stronger predictor of arithmetic ability of the child than of language skills of the child, are valuable extensions of the reference hypothesis H_1, which states that the relation between the arithmetic ability of the father and the mother on the one hand and the arithmetic ability of the child on the other hand should be positive.

TABLE 6.11: Description of Populations in Agreement with H_1 and H_2

Population in Agreement with H_1				
Age	8	11	14	17
Means Girls	.1	.233	.366	.5
Means Boys	0	0	0	0
Population in Agreement with H_2				
Age	8	11	14	17
Means Girls	.5	.366	.233	.1
Means Boys	0	0	0	0
Within Group Covariance Matrix				
Age	8	11	14	17
8	1	.3	.3	.3
11	.3	1	.3	.3
14	.3	.3	1	.3
17	.3	.3	.3	1

6.6   Repeated Measures Analysis

6.6.1   Hypotheses

In Section 2.6, a hypothetical researcher was interested in the development of depression over time (measured at the ages of 8, 11, 14, and 17) of girls and boys. As was previously elaborated in Section 5.10.2, the reliability of the Bayes factor increases with the specificity of the hypotheses under consideration. Our hypothetical researcher has two specific competing expectations with respect to the development of depression that will be used to further illustrate this point:

$$H_1 : \begin{array}{c} \mu_{8g} < \mu_{11g} < \mu_{14g} < \mu_{17g} \\ |\mu_{8b} - \mu_{11b}| < .7 \\ |\mu_{11b} - \mu_{14b}| < .7 \\ |\mu_{14b} - \mu_{17b}| < .7 \\ \mu_{8g} > \mu_{8b} \\ \mu_{11g} > \mu_{11b} \\ \mu_{14g} > \mu_{14b} \\ \mu_{17g} > \mu_{17b} \end{array} , \qquad (6.10)$$

and

$$H_2 : \begin{array}{c} \mu_{8g} > \mu_{8b} \\ \mu_{8g} - \mu_{8b} > \mu_{11g} - \mu_{11b} > \mu_{14g} - \mu_{14b} > \mu_{17g} - \mu_{17b} \\ |\mu_{8b} - \mu_{11b}| < .7 \\ |\mu_{11b} - \mu_{14b}| < .7 \\ |\mu_{14b} - \mu_{17b}| < .7 \end{array} . \qquad (6.11)$$

Note that μ_{8g} denotes the population mean of the depression level of 8-year-old girls, similarly, μ_{14b} denotes the depression level of 14-year-old boys.

6.6.2   Sample Size Determination

The parameters for two populations, one in agreement with H_1 and one in agreement with H_2, can be found in Table 6.11. Note that in each population the size of the effect $e = |\mu_{8g} - \mu_{17g}|/\sigma_{y_{8g} - y_{17g}} = |.1 - .5|/\sqrt{(1 + 1 - 2 \times .3)} = .34$. In each cell of the design,

TABLE 6.12: Error Probabilities for Repeated Measures Analysis

Population	BF_{1a}	BF_{2a}	BF_{12}
$N = 20$			
H_1	$4\% < 1$	$23\% > 1$	$2\% < 1$
H_2	$47\% > 1$	$5\% < 1$	$14\% > 1$
$N = 40$			
H_1	$0\% < 1$	$18\% > 1$	$0\% < 1$
H_2	$36\% > 1$	$1\% < 1$	$5\% > 1$

1,000 data matrices are sampled from the corresponding population for $N = 20$ and $N = 40$ per group, respectively. In Appendix 6D the `GenMVLData` command file used is presented. After sampling 1,000 data matrices, `BIEMS` can be used to compute BF_{1a} and BF_{2a} for each data matrix. The resulting error probabilities are displayed in Table 6.12 and render the following decision rules:

- Prefer H_1 over H_2 if $BF_{12} > 1$.

- Prefer H_2 over H_1 if $BF_{12} < 1$.

Note that due the small error probabilities that are observed for BF_{12} (loosely ignoring the fact that 14% is larger than 10%), an indecision interval is not necessary. Note also that the error probabilities for BF_{1a} and BF_{2a} are substantially larger than the error probabilities for BF_{12}. This further illustrates that the reliability of the Bayes factor increases with the specificity of the hypotheses under investigation. Although the study in Table 6.12 is rather limited, it does provide valuable information with respect to the sample sizes needed for a reliable evaluation of H_1 and H_2: between 20 and 40 persons per group are sufficient if the populations of interest resemble the populations displayed in Table 6.11.

6.6.3   Hypothesis Evaluation Using `BIEMS`

After an inspection of the sample size determination in the previous section, it was decided to use a sample of 20 girls and 20 boys. The data in Table 6.13 were previously presented in Section 2.6. For analysis with `BIEMS`, the data must be placed in a text file named `data.txt`. The first four columns contain the scores on depression at ages 8, 11, 14, and 17, respectively. The last column states that the girls belong to group 1 and the boys to group 2.

The `BIEMS` command files presented in Appendix 6D rendered $BF_{12} = BF_{1a}/BF_{2a} = 3.07/11.18 = .27$. Applying the decision rules it can straightforwardly be concluded that H_2 is preferred over H_1. Note that $BF_{2a} = 11.18$. This implies that the constraints of H_2 are supported by the data and, consequently, that H_2 is *not* the best of two ill-fitting hypotheses (see also Property 4.1 in Chapter 4). The conditional error probabilities, that is, posterior model probabilities corresponding to $BF_{12} = .47$ are .21 and .79 for H_1 and H_2, respectively. This implies that there is some but not yet convincing evidence in favor of H_2. This conclusion is supported by the sample means that are displayed in Table 6.14.

TABLE 6.13: Example Data for Repeated Measures Analysis

Depression at Age					Depression at Age				
8	11	14	17	Gender	8	11	14	17	Gender
8	10	12	8	1	16	4	15	16	2
11	8	5	13	1	4	20	4	9	2
15	22	16	19	1	2	5	9	5	2
6	12	13	12	1	8	19	1	13	2
9	14	5	10	1	12	7	3	6	2
14	12	5	5	1	12	17	22	8	2
3	11	19	6	1	13	15	17	9	2
12	8	11	5	1	6	4	7	11	2
5	14	8	6	1	0	6	1	3	2
8	0	7	11	1	15	4	11	6	2
5	12	14	11	1	13	3	12	12	2
10	12	11	8	1	7	12	8	2	2
19	9	11	8	1	4	7	9	5	2
9	16	17	1	1	13	12	4	9	2
13	19	10	13	1	9	12	3	8	2
15	19	13	12	1	11	8	19	16	2
20	16	16	12	1	7	8	7	10	2
16	3	6	12	1	7	11	13	10	2
13	13	13	16	1	9	12	12	7	2
5	13	9	10	1	9	18	11	7	2

6.7   Discussion

6.7.1   Sample Size Determination

Chapters 4, 5, and 6 provided examples of Bayesian evaluation of informative hypotheses. In its most basic form, this approach consists of the following steps. First of all, the informative hypothesis of interest must be formulated. Subsequently, the hypotheses to which the hypothesis of interest will be compared must be determined. Options are the null hypothesis, the complement of the informative hypotheses, the unconstrained counterpart of the informative hypothesis and competing informative hypotheses. Finally, the software package BIEMS (see Mulder, Hoijtink, and de Leeuw (In Press) and Section 9.2.2) can be used to obtain the Bayes factors and Posterior Model Probabilities with which the hypotheses under consideration can be evaluated. Examples of hypotheses, data, and evaluation using BIEMS were provided for ANOVA (Chapters 4 and 5), ANCOVA and multiple regression (Chapter 5), and multivariate normal linear models (Chapter 6).

The approach sketched in the previous paragraph can be elaborated with a step in which the effect of the sample sizes that are considered by a researcher on the reliability of the

TABLE 6.14: Sample Means and Standard Errors for the Data in Table 6.13

Group		Age 8	Age 11	Age 14	Age 17
Girls	Mean	10.80	12.15	11.05	9.90
	SE	1.08	1.16	.94	.93
Boys	Mean	8.85	10.20	9.40	8.60
	SE	.97	1.21	1.33	.84

resulting Bayes factors can be determined. To execute this step, the following ingredients are necessary: the hypotheses under consideration; for each hypothesis, one or more specifications of populations representing the effect sizes that are of interest; and the sample sizes that are of interest. Subsequently, `GenMVLData` (see Section 9.2.4) can be used to generate/simulate data from the populations of interest, and `BIEMS` can be used to analyze these data. Two different approaches to sample size determination have been presented: a simple and easy-to-execute approach that renders an *impression* of the sample sizes that are needed in order to obtain reliable Bayes factors (Chapter 5), and a time-consuming approach that will render actual error probabilities and enable researchers to specify an interval of indecision (Chapter 6). Note that the simple and easy-to-execute approach is facilitated by the `BIEMS Windows` user interface in which data generation and analysis are nicely integrated.

Adding a step in which sample size determination is considered for Bayesian evaluation of informative hypotheses is an option that not all researchers will use. Although there will rarely be a one-to-one fit between the situations covered in Chapters 5 and 6 and the situation of interest to a researcher, there may be enough similarities to obtain a good indication of the sample sizes needed for the situation at hand. If this is not the case, researchers are currently well advised to use the approach presented in Chapter 5. Although the approach presented in Chapter 6 is preferred, it requires a lot of patience because the generation and analysis of 1,000 data matrices simulated for each cell in a simulation design is rather time consuming. On a personal computer that in 2010 was considered "middle of the road," it took less that a week to create Table 6.12 presented in Section 6.6, but about 3 months to create Table 6.9 presented in Section 6.5.

6.7.2 Bayesian Evaluation of Informative Hypotheses

This chapter will conclude with a discussion of features of Bayesian evaluation of informative hypotheses that became highlighted in the context of sample size determination.

Principle 5.1 presented in Chapter 5 stated that 1 is a natural benchmark for the evaluation of a Bayes factor. If $BF_{mm'} > 1$, H_m is preferred by the data; and if $BF_{mm'} < 1$, $H_{m'}$ is preferred. As was shown in Chapter 6, the natural benchmark is not always a reliable benchmark. The interval of indecision was introduced. This interval states that values of the Bayes factor in the range L to U cannot be used to reliably distinguish H_m from $H_{m'}$. What could be observed in Chapter 6 is that the interval of indecision becomes smaller if the sample size is increased. In fact, if the sample size becomes large enough, the interval of indecision will disappear, which implies that the benchmark of 1 is not only natural, but also reliable.

As could be seen in both Chapters 5 and 6, the sample sizes needed in order to be able to reliably evaluate informative hypotheses are within reach of most behavioral and social scientists. In the context of ANOVA and ANCOVA, samples of 10 to 20 persons per group were usually sufficient. For the multiple regression examples presented in Chapter 5, samples of 20 to 40 persons were usually sufficient. For the examples in the context of the multivariate normal linear model presented in Chapter 6, sample sizes of 25 to 50 appear to be sufficient for multivariate one-sided testing and multivariate treatment evaluation, samples of 100 for the evaluation of multiple components in the context of multivariate regression, and samples of 20 to 40 for the evaluation of competing hypotheses in the context of repeated measurements.

In Section 5.2.1, classical power analysis was introduced. Classical power analysis focuses on a dichotomous decision: reject the null hypothesis or not. The error probabilities of interest are incorrectly rejecting the null (the error of the first kind, usually set at 5%), and incorrectly not rejecting the null (the error of the second kind, usually set at 20%). Sample size determination in a Bayesian context as introduced in this book differs in two ways. First of all, the hypotheses under consideration are treated equally, that is, the errors of incorrectly rejecting H_m and $H_{m'}$ are the same. In this book the subjective decision of the author was to set both error probabilities at 10%. Furthermore, if the sample size is not large enough, the Bayesian approach does not enforce a dichotomous decision. It can very well be concluded that the size of a Bayes factor is such that a reliable choice between H_m and $H_{m'}$ is not possible.

6.8   Appendix 6A: GenMVLData and BIEMS, Multivariate One-Sided Testing

In Table 6.15 the GenMVLData command file input.txt used to sample the data matrices used to construct Table 6.1 and Table 6.2 is displayed. As can be seen exact=0, meaning that a data matrix is sampled instead of constructed such that the sample means and variances are identical to their population counterparts. Because there are #DV=3 dependent variables, the standard deviation of each of them must be provided: 1 1 1 and the correlation matrix of the error terms must be provided. Using population means of .3 for each dependent variable (see the string .3 .3 .3), a population effect size of $e = .3/1 = .3$ is realized for each of the three dependent variables.

In Table 6.16 the BIEMS command files used to obtain BF_{1a}/BF_{11_c} and BF_{0a}, respectively, for the data in Table 6.3 are displayed. Note that there are #DV=3 dependent variables; that the sample size N=20; that there are #inequal=3 inequality and #equal=3 equality constraints, respectively; that the dependent variable is not standardized Z(DV)=0; and that the string 1 0 0 followed by 0 in the top panel denotes the first constraint of H_1, that is, $1 \times \mu_1 + 0 \times \mu_2 + 0 \times \mu_3 > 0$.

TABLE 6.15: input.txt for Generating the Multivariate Data for $\rho = .3$, $N = 25$, and $e = .3$ in Tables 6.1 and 6.2

```
#DV #groups #cov iseed exact
3    1       0    -1    0

group sizes
25

standard deviations for the error terms
1 1 1

correlation matrix for the error terms
 1 .3 .3
.3  1 .3
.3 .3  1

parameter matrix
.3 .3 .3

means for normal covariates per group

covariance matrices for normal covariates per group
```

TABLE 6.16: BIEMS Command Files for Multivariate One-Sided Testing Using the Data in Table 6.3

Computation of BF_{1a} and BF_{11_c}					
input_BIEMS.txt					
Input 1:	#DV	#cov	#Tcov	N	iseed
	3	0	0	20	1111
Input 2:	#inequal	#equal	#restr		
	3	0	-1		
Input 3:	sample size	maxBF steps	scale		
	-1	-1	-1		
Input 4:	Z(DV)	Z(IV)			
	0	0			
Input 5:	Z(IV)				

inequality_constraints.txt	
R_m	r_m
1 0 0	0
0 1 0	0
0 0 1	0

Computation of BF_{0a}					
input_BIEMS.txt					
Input 1:	#DV	#cov	#Tcov	N	iseed
	3	0	0	20	1111
Input 2:	#inequal	#equal	#restr		
	0	3	-1		
Input 3:	sample size	maxBF steps	scale		
	-1	-1	-1		
Input 4:	Z(DV)	Z(IV)			
	0	0			
Input 5:	Z(IV)				

equality_constraints.txt	
S_m	s_m
1 0 0	0
0 1 0	0
0 0 1	0

6.9 Appendix 6B: GenMVLData and BIEMS, Multivariate Treatment Evaluation

In Table 6.17 the GenMVLData command file input.txt used to sample the data matrices used to construct Tables 6.5 and 6.6 is displayed. As can be seen, there are 25 persons in each of two groups 25 25; and using population means of .3 for each dependent variable (see the string .3 .3 .3) in the first group and 0 (see the string 0 0 0) in the second group, combined with within group standard deviations of 1 (see the string 1 1 1) a population effect size of $e = (.3 - 0)/1 = .3$ is realized for each of the three dependent variables.

In Table 6.18 the BIEMS command files used to obtain BF_{1a}/BF_{11_c} and BF_{0a}, respectively, for the data in Tables 6.3 and 6.7 are displayed. Note that the combined sample size in both groups is N=40, and, that the string 1 -1 0 0 0 0 followed by 0 in the top panel denotes the first constraint of H_1, that is, $1 \times \mu_{11} - 1 \times \mu_{12} + 0 \times \mu_{21} + 0 \times \mu_{22} \mid 0 \times \mu_{31} + 0 \times \mu_{33} > 0$. The corresponding strings in the bottom panel denote the first constraint of H_0, that is, $1 \times \mu_{11} - 1 \times \mu_{12} + 0 \times \mu_{21} + 0 \times \mu_{22} + 0 \times \mu_{31} + 0 \times \mu_{33} = 0$.

TABLE 6.17: input.txt for Generating the Data for $\rho = .3$, $N = 25$ per group, and $e = .3$ in Tables 6.5 and 6.6

```
#DV #groups #cov iseed exact
3    2       0    -1    0

group sizes
25 25

standard deviations for the error terms
1 1 1

correlation matrix for the error terms
 1 .3 .3
.3  1 .3
.3 .3  1

parameter matrix
.3 .3 .3
 0  0  0

means for normal covariates per group

covariance matrices for normal covariates per group
```

TABLE 6.18: BIEMS Command Files for Multivariate Treatment Evaluation Using the Data in Tables 6.3 and 6.7

Computation of BF_{1a} and BF_{11_c}					
input_BIEMS.txt					
Input 1:	#DV	#cov	#Tcov	N	iseed
	3	0	0	40	1111
Input 2:	#inequal	#equal	#restr		
	3	0	-1		
Input 3:	sample size	maxBF steps	scale		
	-1	-1	-1		
Input 4:	Z(DV)	Z(IV)			
	0	0			
Input 5:	Z(IV)				

inequality_constraints.txt	
R_m	r_m
1 -1 0 0 0 0	0
0 0 1 -1 0 0	0
0 0 0 0 1 -1	0

Computation of BF_{0a}					
input_BIEMS.txt					
Input 1:	#DV	#cov	#Tcov	N	iseed
	3	0	0	40	1111
Input 2:	#inequal	#equal	#restr		
	0	3	-1		
Input 3:	sample size	maxBF steps	scale		
	-1	-1	-1		
Input 4:	Z(DV)	Z(IV)			
	0	0			
Input 5:	Z(IV)				

equality_constraints.txt	
S_m	s_m
1 -1 0 0 0 0	0
0 0 1 -1 0 0	0
0 0 0 0 1 -1	0

6.10 Appendix 6C: `GenMVLData` and `BIEMS`, Multivariate Regression

In Table 6.19 the `GenMVLData` command file `input.txt` used to sample the data matrices used to construct Table 6.9 is displayed. As can be seen, there are 100 persons in the samples 100; and the proportion of variance R^2 explained for the first and second dependent variable are $1 - 7746^2 = .40$ and $1 - .8718^2 = .24$, respectively. Note that .7746 .8718 denotes the residual standard deviations of the first and second dependent variable, respectively. All `#cov=4` predictors are standardized. The next to last entry in Table 6.19 denotes that the mean of each predictor is 0. The last entry in Table 6.19 denotes that the variance of each predictor is 1, and the correlations between the predictors .3. The intercept and regression coefficients relating the predictors to the first dependent variable are 0 .30 .15 .30 .15.

In Table 6.20 the `BIEMS` command files used to obtain $BF_{1\&2,a}$ and $BF_{1,a}$ for the data displayed in Appendix 2A are displayed. Note that both the dependent variable `Z(DV)=1` and the predictors `Z(IV)=1` are standardized. Note also that the last constraint in the bottom panel denotes that $0 \times \beta_{A0} + 0 \times \beta_{A1} + 0 \times \beta_{A2} + 0 \times \beta_{A3} + 0 \times \beta_{A4} + 0 \times \beta_{L0} + 0 \times \beta_{L1} + 0 \times \beta_{L2} + 0 \times \beta_{L3} + 1 \times \beta_{L4} > 0$, that is, it is the last constraint of H_1. Note that this listing first addresses the regression coefficients relating the predictors to the first dependent variable (arithmetic skills) and subsequently the regression coefficients relating the predictors to the second dependent variable (language skills).

If `BIEMS` is executed using the `Windows` user interface, by default the unconstrained prior distribution will be chosen such that it is compatible for all hypotheses under consideration. If the stand alone version of `BIEMS` is used, compatibility has to be checked, and if necessary, enforced by the user. How this can be done is elaborated in Section 9.2.2.1. Note that `#restr=8` both in the top panel for the computation of $BF_{1\&2,a}$ and in the bottom panel for $BF_{1,a}$. The command `#restr=-1` means that `BIEMS` will use default constraints to construct the unconstrained prior distribution of the regression coefficients. However, the default unconstrained prior of $H_{1\&2}$ is incompatible with the default unconstrained prior of H_1. This can be confirmed using the guidelines formulated in Section 4.4.2: if the inequality constraints in $H_{1\&2}$ and H_1 are replaced by equality constraints, the resulting hypotheses are *not* the same.

To ensure that compatible unconstrained priors are used, they must be specified manually. The command `#restr=8` states that eight constraints on the prior means of the regression coefficients can be found in a text file with the name `restriction_matrix.txt`. The matrix used for the example at hand can be found in Table 6.21. It is constructed using two rules that will be elaborated in Section 9.2.2.1:

- Set the prior means of all the parameters involved in the informative hypotheses equal to each other. As can be seen, in the first row of Table 6.21 it is specified that the prior mean of β_{A1} is equal to the prior mean of β_{A2}. Six more lines are needed to specify that the eight regression coefficients all have the same prior mean.

- If one or more of the parameters are constrained to be larger or smaller than 0 in at least one of the informative hypotheses under consideration (as can be seen in Table 6.20, β_{A1} is constrained to be larger than 0 in both $H_{1\&2}$ and H_1), the prior mean of one of them must be set equal to 0. As can be seen in the last line of Table 6.21, the prior mean of β_{L4} is set equal to 0.

As can be seen, the restrictions in both H_1 and $H_{1\&2}$ with the inequality constraints replaced by equality constraints are implied by the imaginary hypothesis $H_* : \beta_{A1} = \ldots = \beta_{L4} = 0$ corresponding to the restrictions in Table 6.21.

TABLE 6.19: `input.txt` for Generating the Data for H_1 through H_4 in Table 6.9

```
#DV #groups #cov iseed exact
2   1       4    -1    0

group sizes
100

standard deviations for the error terms
.7746 .8718

correlation matrix for the error terms
 1 .3
.3  1

parameter matrix
0    0
.30 -.17
.15  .34
.30 -.17
.15  .34

means for normal covariates per group
0
0
0
0

covariance matrices for normal covariates per group
 1 .3 .3 .3
.3  1 .3 .3
.3 .3  1 .3
.3 .3 .3  1
```

TABLE 6.20: BIEMS Command Files for Multivariate Regression: Comparing $H_{1\&2}$ with H_1 Using the Data in Appendix 2A

	Computation of $BF_{1\&2,a}$					
input_BIEMS.txt						
Input 1:	#DV	#cov	#Tcov	N	iseed	
	2	4	0	100	1111	
Input 2:	#inequal	#equal	#restr			
	8	0	8			
Input 3:	sample size	maxBF steps	scale			
	-1	-1	-1			
Input 4:	Z(DV)	Z(IV)				
	1	1				
Input 5:	Z(IV)					

inequality_constraints.txt

R_m	r_m
0 1 0 0 0 0 0 0 0 0	0
0 0 0 1 0 0 0 0 0 0	0
0 0 0 0 0 0 0 1 0 0	0
0 0 0 0 0 0 0 0 0 1	0
0 1 -1 0 0 0 0 0 0 0	0
0 0 0 1 -1 0 0 0 0 0	0
0 0 0 0 0 0 -1 1 0 0	0
0 0 0 0 0 0 0 0 -1 1	0

	Computation of BF_{1a}					
input_BIEMS.txt						
Input 1:	#DV	#cov	#Tcov	N	iseed	
	2	4	0	100	1111	
Input 2:	#inequal	#equal	#restr			
	4	0	8			
Input 3:	sample size	maxBF steps	scale			
	-1	-1	-1			
Input 4:	Z(DV)	Z(IV)				
	1	1				
Input 5:	Z(IV)					

equality_constraints.txt

R_m	r_m
0 1 0 0 0 0 0 0 0 0	0
0 0 0 1 0 0 0 0 0 0	0
0 0 0 0 0 0 0 1 0 0	0
0 0 0 0 0 0 0 0 0 1	0

TABLE 6.21: BIEMS Command File `restriction_matrix.txt` Used for the Analysis of the Data in Appendix 2A

0	1	-1	0	0	0	0	0	0	0	0
0	0	1	-1	0	0	0	0	0	0	0
0	0	0	1	-1	0	0	0	0	0	0
0	0	0	0	1	0	-1	0	0	0	0
0	0	0	0	0	0	1	-1	0	0	0
0	0	0	0	0	0	0	1	-1	0	0
0	0	0	0	0	0	0	0	1	-1	0
0	0	0	0	0	0	0	0	0	1	0

6.11   Appendix 6D: GenMVLData and BIEMS, Repeated Measures

In Table 6.22 the GenMVLData command file input.txt used to sample the data matrices needed for the construction of Table 6.12 is displayed.

In Table 6.23 the BIEMS command files used to obtain BF_{1a} and BF_{2a} for the data in Table 6.13 are displayed. Note that the combined sample size in both groups is N=40, and that #inequal=13 inequality constraints are used to construct H_1 and #inequal=10 to construct H_2. Note furthermore that sample size=100,000. As will be elaborated in Section 9.2.2.3, the BIEMS output showed that the estimates of BF_{1a} and BF_{2a} obtained using the default command sample size=-1 were inaccurate. Therefore the default sample size from prior and posterior is increased from the default value of 20,000 to 100,000. The string -1 0 1 0 0 0 0 0 followed by 0 in the top panel states that $-1 \times \mu_{8g} + 0 \times \mu_{8b} + 1 \times \mu_{11g} + 0 \times \mu_{11b} + 0 \times \mu_{14g} + 0 \times \mu_{14b} + 0 \times \mu_{17g} + 0 \times \mu_{17b} > 0$, that is, $\mu_{8g} < \mu_{11g}$, which is the first constraint appearing in H_1. Note that the last two strings in the bottom panel jointly state that $|\mu_{14b} - \mu_{17b}| < .7$. Note finally that the default unconstrained prior distributions corresponding to H_1 and H_2 are identical and thus compatible. This can be verified using the output rendered by BIEMS when BF_{1a} and BF_{2a} are computed. The default restriction_matrix.txt used is printed at the bottom of each output. It is the same for both hypotheses.

TABLE 6.22: input.txt for Generating the Data for H_1 in Table 6.12

```
#DV #groups #cov iseed exact
4    2      0    -1     0

group sizes
20 20

standard deviations for the error terms
1 1 1 1

correlation matrix for the error terms
 1 .3 .3 .3
.3  1 .3 .3
.3 .3  1 .3
.3 .3 .3  1

parameter matrix
.1 .233 .366 .5
0  0    0    0

means for normal covariates per group

covariance matrices for normal covariates per group
```

TABLE 6.23: BIEMS Command Files for Repeated Measures Analysis Using the Data in Table 6.13

		Computation of BF_{1a}				
input_BIEMS.txt						
Input 1:		#DV	#cov	#Tcov	N	iseed
		4	0	0	40	1111
Input 2:		#inequal	#equal	#restr		
		13	0	-1		
Input 3:		sample size	maxBF steps	scale		
		100000	-1	-1		
Input 4:		Z(DV)	Z(IV)			
		0	0			
Input 5:		Z(IV)				

inequality_constraints.txt

R_m								r_m
-1	0	1	0	0	0	0	0	0
0	0	-1	0	1	0	0	0	0
0	0	0	0	-1	0	1	0	0
0	1	0	-1	0	0	0	0	-.7
0	-1	0	1	0	0	0	0	-.7
0	0	0	1	0	-1	0	0	-.7
0	0	0	-1	0	1	0	0	-.7
0	0	0	0	0	1	0	-1	-.7
0	0	0	0	0	-1	0	1	-.7
1	-1	0	0	0	0	0	0	0
0	0	1	-1	0	0	0	0	0
0	0	0	0	1	-1	0	0	0
0	0	0	0	0	0	1	-1	0

		Computation of BF_{2a}				
input_BIEMS.txt						
Input 1:		#DV	#cov	#Tcov	N	iseed
		4	0	0	40	1111
Input 2:		#inequal	#equal	#restr		
		10	0	-1		
Input 3:		sample size	maxBF steps	scale		
		100000	-1	-1		
Input 4:		Z(DV)	Z(IV)			
		0	0			
Input 5:		Z(IV)				

equality_constraints.txt

S_m								s_m
1	-1	0	0	0	0	0	0	0
1	-1	-1	1	0	0	0	0	0
0	0	1	-1	-1	1	0	0	0
0	0	0	0	1	-1	-1	1	0
0	1	0	-1	0	0	0	0	-.7
0	-1	0	1	0	0	0	0	-.7
0	0	0	1	0	-1	0	0	-.7
0	0	0	-1	0	1	0	0	-.7
0	0	0	0	0	1	0	-1	-.7
0	0	0	0	0	-1	0	1	-.7

Part III

Other Models, Other Approaches, and Software

This part of the book consists of Chapters 7, 8, and 9. In Chapter 7 Bayesian evaluation of informative hypotheses is discussed for contingency tables, multilevel models, latent class models, and statistical models in general. Chapter 8 presents non-Bayesian approaches for the evaluation of informative hypotheses for which software exists with which these approaches can be implemented. Specifically, an approach based on null hypothesis significance testing and the order-restricted information criterion are discussed. Chapter 9 presents an overview of the software that is available for the evaluation of informative hypotheses.

Symbol Description

	Chapter 7	
	Contingency Tables	
θ	Odds ratio.	
π_{ab}	Probability of category a of a first and category b of a second categorical variable.	
	Multilevel Models	
β_{0i}	Random intercept for the i-th person.	
β_{1i}	Random slope for the i-th person.	
t	Time point.	
y_{ti}	Weight of person i at time point t.	
ϵ_{ti}	Residual of person i for the t-th measurement of weight.	
μ_0	Average of the random intercepts of all persons.	
μ_1	Average of the random slopes of all persons.	
ϕ_{00}	Variance of the random intercepts.	
ϕ_{11}	Variance of the random slopes.	
ϕ_{01}	Covariance of the random intercepts and slopes.	
	Latent Class Models	
π_{dj}	Probability of responding correctly to item j for members of class d.	

ω_d	Probability of belonging to class d.
	Statistical Models in General
$g(\cdot)$	Posterior distribution.
$h(\cdot)$	Prior distribution.
$f(\cdot)$	Density of the data.
$\boldsymbol{Y}$	Observed data.
$\boldsymbol{X}$	Ancillary data.
$\boldsymbol{\theta}$	Parameters subjected to (in)equality constraints.
$\boldsymbol{\phi}$	Nuisance parameters.
$\boldsymbol{R}_m$	Matrix containing inequality constraints.
$\boldsymbol{r}_m$	Vector containing effect sizes for inequality constraints.
$t(\boldsymbol{\theta})$	Monotone transformation applied to each element of $\boldsymbol{\theta}$ or to groups of elements of the same size.
	Chapter 8
$\boldsymbol{S}_m$	Matrix containing equality constraints.
$\boldsymbol{s}_m$	Vector containing effect sizes for equality constraints.
p_m	Penalty for the complexity of hypothesis H_m.

Chapter 7

Beyond the Multivariate Normal Linear Model

7.1 Introduction

In the previous chapters the evaluation of informative hypotheses in the context of the univariate and multivariate normal linear model was elaborated. In this chapter the application of informative hypotheses is discussed and illustrated in the context of contingency tables, multilevel models, and latent class analysis. Furthermore, a general framework for

TABLE 7.1: A Hypothetical Contingency Table (Entries are Number of Persons)

		Age	
Attitude	1:"<25"	2:"25–67"	3:">67"
1:Left-Wing	32	45	15
2:Right-Wing	17	55	11

the evaluation of informative hypotheses is provided. For each context the following topics are discussed:

- Using a concrete example (thus sacrificing some generality but keeping matters rather accessible), the statistical model at hand is introduced. Because informative hypotheses are constructed via constraints on the parameters of a statistical model (in the context of the univariate normal linear model, these were means and regression coefficients), there will be explicit attention for the meaning and interpretation of these parameters.

- Informative hypotheses for the example data at hand are presented and evaluated. Illustrations of the software packages and command files used for these evaluations are presented and discussed.

- Specific features of the evaluation of informative hypotheses for the model at hand are discussed.

- References are given to papers containing the statistical foundations, and papers containing applications of the evaluation of informative hypotheses for the model at hand.

7.2   Contingency Tables

7.2.1   Model and Parameters

The interested reader is referred to Agresti (2007) for a general introduction to the analysis of categorical data and contingency tables. A contingency table is the joint representation of the responses of persons to two or more categorical variables. A categorical variable consists of a number of categories, each in a sample of persons is assigned to one of these categories. Examples of categorical variables are Gender (male or female), Age (underage, adult, retired), and Occupation (blue collar, white collar, and unemployed). A hypothetical contingency table is displayed in Table 7.1. For each of 175 persons the political *Attitude* (left-wing or right-wing) is crossed with the *Age* level (younger than 25, between 25 and 67, or older than 67). Note that the numbers (1–3 for *Age* and 1–2 for *Attitude*) in front of the category labels are category numbers.

In Table 7.2 the parameters of the model for the data in Table 7.1 are displayed. Each parameter is the probability that a person randomly sampled from the population of interest belongs to the combination of categories at hand. For example, π_{21} is the probability that a person is between 25 and 67 years old, and has a left-wing political *Attitude*. In Table 7.3 estimates of the population probabilities obtained using the data in Table 7.1 are displayed. Note that the sum of these probabilities is equal to 1.0.

TABLE 7.2: Model Parameters

Attitude	1: "<25"	2: "25–67"	3: ">67"
	Age		
1:Left-Wing	π_{11}	π_{21}	π_{31}
2:Right-Wing	π_{12}	π_{22}	π_{32}

7.2.2 Hypothesis Evaluation Using `ContingencyTable`

Two types of elements that are often useful when constructing informative hypotheses for data summarized in contingency tables are presented. The first type is the oddsratio:

$$\theta_{ab} = \frac{\pi_{ab} \times \pi_{a+1,b+1}}{\pi_{a+1,b} \times \pi_{a,b+1}}, \tag{7.1}$$

where, referring to Table 7.2, a can obtain the values 1 and 2, and b can only obtain the value 1. Using the numbers in Table 7.3, estimates of two odds ratio can be computed: $\hat{\theta}_{11} = \frac{.18 \times .31}{.26 \times .10} = 2.15$ and $\hat{\theta}_{21} = \frac{.26 \times .06}{.09 \times .31} = 0.56$. Odds ratios can be interpreted relative to the number 1. The odds ratio of 2.15 is larger than 1, which implies that there is a positive relation between the categories 1 and 2 of *Age* and *Attitude*. This means that persons scoring in the higher category of *Age* (that is, 2 instead of 1) also tend to score in the higher category of *Attitude* (that is, 2 instead of 1). Stated otherwise, with increasing *Age*, persons more frequently have a right-wing political *Attitude*. The odds ratio of 0.56 (which is smaller than 1) implies that there is a negative relation between the categories 2 and 3 of *Age* and 1 and 2 of *Attitude*. Persons scoring in the higher category of *Age* (that is, 3 instead of 2) tend to score in the lower category of *Attitude* (that is, 1 instead of 2). Stated otherwise, with a further increase of *Age* persons more frequently have a left-wing political *Attitude*.

The second type are probabilities that are weighted by the sum of the probabilities in the corresponding row or column. Consider, for example,

$$\frac{\pi_{11}}{\pi_{11} + \pi_{12}}, \frac{\pi_{21}}{\pi_{21} + \pi_{22}}, \text{and } \frac{\pi_{31}}{\pi_{31} + \pi_{32}}. \tag{7.2}$$

Here each probability in the first row of Table 7.2 is weighted by the sum of the probabilities in the corresponding column. The result is the proportion persons with a left-wing *Attitude* for each *Age* level. Estimates of these proportions are: $\frac{.18}{.18+.10} = .64$, $\frac{.26}{.26+.31} = .46$, and $\frac{.09}{.09+.06} = .60$, respectively, for increasing *Age* levels.

Using elements like (7.1) and (7.2), informative hypotheses for contingency tables can be constructed. If the 175 persons in the hypothetical sample all have an academic background, an expectation with respect to the relation between *Age* and *Attitude* could be the following. Most persons start with a left-wing political *Attitude*. However, if later in life they reap the benefits of their academic education, have mortgages and other financial obligations, the *Attitude* may become more right-wing. Still later in life when they are financially secure, they may move back to the left-wing. There are various ways in which this expectation can

TABLE 7.3: Estimates of the Population Probabilities

Attitude	1: "<25"	2: "25–67"	3: ">67"
	Age		
1:Left-Wing	.18	.26	.09
2:Right-Wing	.10	.31	.06

TABLE 7.4: `ContingencyTable` Data File `data.txt`

Age	Attitude	Number of Persons
1	1	32
2	1	45
3	1	15
1	2	17
2	2	55
3	2	11

be translated into an informative hypothesis. Consider, for example,

$$H_1: \quad \frac{\pi_{11}/\pi_{12} > 1}{\frac{\pi_{11}}{\pi_{11}+\pi_{12}} > \frac{\pi_{21}}{\pi_{21}+\pi_{22}} < \frac{\pi_{31}}{\pi_{31}+\pi_{32}}} \ . \tag{7.3}$$

The first element in this hypothesis states that most persons younger than 25 have a left-wing political *Attitude*. The second element states that the proportion of left-wingers decreases from younger than 25 to 25–67, and increases from 25–67 to 67+. Another possibility is

$$H_{1'}: \quad \begin{array}{c} \pi_{11}/\pi_{12} > 1 \\ \theta_{11} > 1 \\ \theta_{21} < 1 \end{array} \ . \tag{7.4}$$

The first element is the same as in the previous hypothesis. The first odds ratio states that there is a positive relationship between the first two categories of *Age* and *Attitude*. This implies that persons move to the right-wing. The second odds ratio states that there is a negative relationship between the last two categories of *Age* and *Attitude*. This implies that persons move to the left-wing. Formally, both hypothesis are identical, that is, each one can be rewritten such that the other is obtained.

A hypothesis to which H_1 could be compared is

$$H_0: \quad \frac{\pi_{11}/\pi_{12} > 1}{\frac{\pi_{11}}{\pi_{11}+\pi_{12}} = \frac{\pi_{21}}{\pi_{21}+\pi_{22}} = \frac{\pi_{31}}{\pi_{31}+\pi_{32}}} \ , \tag{7.5}$$

that is, persons younger than 25 are more likely to be left-wing than right-wing, but the proportion left-wingers does not change with *Age*.

To use the software package `ContingencyTable` (see Section 9.2.5), the data from Table 7.1 must be placed in a text file with the name `data.txt` in the format presented in Table 7.4. The first column gives the category numbers of *Attitude*, the second column the category numbers of *Age*, and the last column the number of observations in each combination of categories. The other commands necessary must be placed in a text file with the name `ini.txt`, as illustrated for the example at hand in Table 7.5. Note that `#dim_crosstab=2` denotes that the contingency table contains the cross-tabulation of two categorical variables, `#ncel_crosstab=6` denotes that there are six cells in the contingency table, and `#n_models=2` denotes that two hypotheses must be evaluated. Under `#model 1`, H_1 is presented. Note that, c1 through c6 correspond to the rows in Table 7.4, for example, c2 denotes π_{21}, that is, the probability of being in *Age* category "25–67" and *Attitude* category "Left-Wing". Note the similarity between H_1 and its specification in Table 7.5. This similarity makes it very easy to specify hypotheses in `ini.txt`. Under `#model 2`, H_0 is presented.

To obtain Bayes factors from `ContingencyTable`, the program should be run a few times. Multiple runs are needed to ensure that stable estimates of Bayes factors are obtained. See

TABLE 7.5: ContingencyTable Command File `ini.txt`

```
## description
#dim_crosstab
2
#ncel_crosstab
6
#n_models
2
#model 1
"c1/c4 > 1"
"c1/(c1+c4)>c2/(c2+c5)"
"c3/(c3+c6)>c2/(c2+c5)"
#model 2
"c1/c4 > 1"
"c1/(c1+c4)=c2/(c2+c5)"
"c3/(c3+c6)=c2/(c2+c5)"
#eof
```

Section 9.2.5 for further elaboration of this topic. Running ContingencyTable four times rendered the following results for $BF_{10} = BF_{1a}/BF_{0a}$: $3.70/2.20 = 1.68$, $3.68/1.51 = 2.44$, $3.74/1.42 = 2.63$, and $3.75/1.58 = 2.37$. As can be seen, the estimates of BF_{0a} are somewhat instable. Using these results it can be concluded that H_1 is about 2.5 times as likely as H_0 after observing the data. Note that the "outlier" 1.68 is ignored and that 2.5 is based on the other "homogeneous" values of BF_{10}. This is some evidence in favor of H_1 over H_0.

7.2.3 Specific Features

Klugkist, Laudy, and Hoijtink (2010) present some results with respect to the reliability of the Bayes factor. They focus on the following hypotheses for a $2 \times 2 \times 2$ contingency table: $H_0 : \theta_1 = \theta_2$, $H_0 : \theta_1 > \theta_2$, and $H_a : \theta_1, \theta_2$, where θ_1 and θ_2 denote odds ratios for the first and second slice of the contingency table, respectively. The method used was similar to the method used in Chapter 5, that is, construct data matrices that are in perfect agreement with the population of interest. Their results show that a sample size of between 200 and 400 persons is needed to obtain reliable Bayes factors. Because their contingency table contained eight cells, a very rough guideline is to use a sample size of 25–50 for each cell in the contingency table.

7.2.4 Further Reading

The unconstrained prior distribution for the cell probabilities in a contingency table is presented in Appendix 10B. A convenient choice is to use a noninformative Dirichlet distribution (see Appendix 10C) with $a_0 = 1$. The unconstrained posterior distribution is proportional to the product of the likelihood presented in Appendix 10A and the unconstrained prior distribution. The evaluation of informative hypotheses in the context of contingency tables is discussed in Laudy and Hoijtink (2007), Klugkist, Laudy, and Hoijtink (2010), and Laudy (2008). Applications are presented in Meeus, van de Schoot, Keijsers, Schwartz, and Branje (2010) and Meeus, van de Schoot, Klimstra, and Branje (2011). Further information with respect to ContingenyTable can be found in Section 9.2.5.

TABLE 7.6: Example Data for Multilevel Analysis

i	$x_1 = -1.5$	$x_2 = -.5$	$x_3 = .5$	$x_4 = 1.5$
1	101	105	110	115
2	140	142	141	143
3	132	140	141	140
4	125	128	133	138
5	160	158	165	166
6	179	178	173	174
7	147	152	156	160
8	143	150	155	155
9	110	112	111	112
10	115	117	121	119
11	120	119	126	122
12	126	130	133	130

7.3   Multilevel Models

7.3.1   Model, Parameters, and Informative Hypotheses

The interested reader is referred to Hox (2010) for an introduction to multilevel modeling. Consider the hypothetical data set in Table 7.6 concerning the weight in pounds of $N = 12$ freshman at $t = 1, \ldots, 4$ time points: at the beginning of the first year in college $x_1 = -1.5$, around Christmas $x_2 = -.5$, in the spring $x_3 = .5$, and at the end of the first year $x_4 = 1.5$. The nested structure of these data (measurements within persons) can be accounted for using a multilevel model:

$$y_{ti} = \beta_{0i} + \beta_{1i}x_t + \epsilon_{ti} \text{ for } t = 1, \ldots, 4, \tag{7.6}$$

where

$$\begin{bmatrix} \beta_{0i} \\ \beta_{1i} \end{bmatrix} \sim \mathcal{N}\left(\begin{bmatrix} \mu_0 \\ \mu_1 \end{bmatrix}, \begin{bmatrix} \phi_{00} & \phi_{01} \\ \phi_{01} & \phi_{11} \end{bmatrix} \right) \tag{7.7}$$

for $i = 1, \ldots, N$, and $\epsilon_{ti} \sim \mathcal{N}(0, \sigma^2)$. The model in (7.6) is a growth curve model that specifies a linear development of weight over time for each freshman. Note that y_{ti} denotes the weight of person i at time point t. As can be seen in Figure 7.1, β_{0i} denotes the average weight of person i in the first year of college. Furthermore, β_{1i} denotes the average increase in weight between two adjacent time points, that is, $3 \times \beta_{1i}$ denotes the increase in weight during the first year of freshman i. The average of β_{0i} over all persons is denoted by μ_0 and the average of β_{1i} over all persons is denoted by μ_1. The variance of β_{0i} and β_{1i} is denoted by ϕ_{00} and ϕ_{11}, respectively, and their covariance by ϕ_{01}. Finally, the residual variance is denoted by σ^2.

The hypotheses of interest is whether the "Freshman 15" exists or not, that is, whether freshman gain a lot of weight in their first year of college. This can, for example, be achieved via the comparison of

$$H_1 : \begin{matrix} \mu_0 > 100 \\ 3 \times \mu_1 > 10 \end{matrix}, \tag{7.8}$$

that is, the average weight of all freshman is larger than 100 pounds and the average increase in weight of all freshman is larger than 10 pounds, with

$$H_2 : \begin{matrix} \mu_0 > 100 \\ 3 \times \mu_1 < 10 \end{matrix}. \tag{7.9}$$

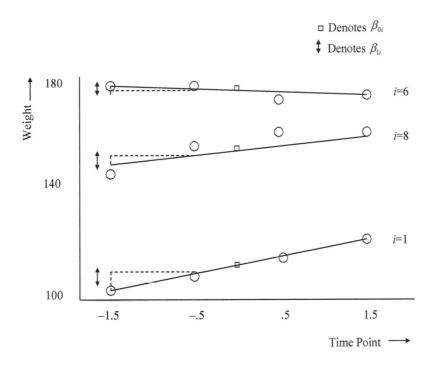

FIGURE 7.1: A visual illustration of the multilevel model (7.6).

7.3.2 Hypothesis Evaluation Using WinBUGS

The software package WinBUGS (see Section 9.2.7) was developed to facilitate Bayesian data analysis. It can be used for the evaluation of informative hypotheses as long as these hypotheses are not specified using equality or about equality constraints. In Appendices 7A and 7B it is elaborated how

$$BF_{12} = BF_{1a}/BF_{2a} = \frac{f_1}{c_1}\bigg/\frac{f_2}{c_2} \tag{7.10}$$

can be obtained using WinBUGS. As is elaborated in Section 7.5.3, complexity and fit can be estimated using a sample from the prior and posterior distribution, respectively. Execution of the code in Appendix 7A renders the proportion of the unconstrained prior distribution in agreement with H_1, that is, $c_1 = .229$. After changing c2 <- step(3 * mu[2]-10) to c2 <- step(10-3 * mu[2]), the corresponding proportion for H_2 is obtained: $c_2 = .231$. Execution of the code in Appendix 7B renders the proportion of the unconstrained posterior distribution in agreement with H_1, that is, $f_1 = .063$. After changing f2 <- step(3 * mu[2]-10) to f2 <- step(10-3 * mu[2]), the corresponding proportion for H_2 is obtained: $f_2 = .929$. Insertion of these numbers in (7.10) renders $BF_{12} = .068$, that is, after observing the data H_2 is about 15 times as likely as H_1, that is, these data do not provide evidence for the "Freshman 15." Note that the code in Appendix 7B can also be used to obtain estimates of μ_0 and μ_1. The 95% central credibility interval for μ_0 was

[124.0,149.5], that is, the assumption that $\mu_0 > 100$ has an excellent fit with the data. The 95% central credibility interval for μ_1 was [0.8,3.55], that is, H_2 has an excellent fit with the data.

As will be elaborated in Section 7.5.3, the precision with which c_m and f_m (and thus BF_{ma}) for $m = 1, 2$ are estimated depends on the size T of the sample from the prior and posterior distribution. As is elaborated in Appendix 7C, `WinBUGS` can be used to compute a 95% central credibility interval for BF_{12} reflecting the uncertainty due to sampling (see Section 10.5.4 for further information). The resulting interval for the example at hand is [.067–.069], which shows that the estimate of BF_{12} obtained using $T = 1,000,000$ is very precise.

7.3.3   Further Reading

The prior distributions used for the multilevel model (7.7) are displayed in Appendix 10B, the density of the data of the multilevel model can be found in Appendix 10A. For both μs in (7.7), a normal prior distribution was specified using a mean of 0 and prior variance $\tau_0^2 \to \infty$. Note that `WinBUGS` requires normal distributions to be specified using precisions instead of variances. Therefore the inverse of the prior variance $I\tau_0^2 \to 0$. The inverse Wishart prior distribution for the covariance matrix ϕ of the βs and the scaled inverse chi-square distribution for the residual variance σ^2 should be specified such that they are vague. As can be seen in Appendix 7B, `WinBUGS` requires the specification of a prior distribution for the inverse of ϕ, that is, $I\phi$, and σ^2, that is, $I\sigma^2$. The corresponding prior distributions are the Wishart (instead of the inverse Wishart) and the gamma (instead of the scaled inverse chi-square). Vague specifications are obtained using $\nu_0 = 2$ and Σ_0 as specified in Appendix 7B (see Appendix 10C for an elaboration of the specification of Σ_0) for the Wishart, and small values, say $a = b = .01$, for the parameters of the gamma distribution (see Appendix 10C).

The evaluation of informative hypotheses in the context of multilevel models is discussed in Kato and Hoijtink (2006) and Kato and Peeters (2008). Further information about `WinBUGS` can be found in Section 9.2.7.

7.4   Latent Class Analysis

7.4.1   Model, Parameters, and Informative Hypotheses

The interested reader is referred to McCutcheon (1987) for an introduction to latent class analysis. Consider the hypothetical data in Table 7.7. They contain the responses of 80 elementary school pupils to the following five questions:

1. $4 + 6 = ?$

2. $32 - 17 = ?$

3. $23 \times 19 = ?$

4. $25/8 = ?$

5. $21 \times 5/25 + 8 = ?$

TABLE 7.7: Example Data for Latent Class Analysis

i	Responses	i	Responses	i	Responses	i	Responses
1	11101	21	10111	41	11111	61	01001
2	11100	22	11111	42	11111	62	01101
3	11010	23	10111	43	01011	63	11111
4	01110	24	11011	44	11110	64	11110
5	11110	25	11111	45	11101	65	11111
6	11111	26	01110	46	10111	66	01100
7	01010	27	10111	47	11111	67	11111
8	11111	28	10110	48	11111	68	11101
9	10011	29	01011	49	11011	69	11101
10	11111	30	11110	50	11111	70	10011
11	10001	31	00000	51	01000	71	00100
12	10000	32	00010	52	00000	72	00100
13	10000	33	11100	53	10011	73	10000
14	00001	34	10001	54	00001	74	00000
15	00100	35	00011	55	00001	75	11001
16	10101	36	00000	56	00101	76	00000
17	00010	37	01010	57	00000	77	00001
18	10000	38	01010	58	10000	78	11000
19	11000	39	10000	59	00000	79	00010
20	00100	40	00100	60	01011	80	00001

which constitute a small test with which arithmetic ability can be measured. Each question can be responded to correctly (the score 1 is given) or incorrectly (the score 0 is given). Latent class models can be used to cluster the pupils in classes. The classes are called latent because it is unknown to to which class each pupil belongs; stated otherwise, the class-membership of each pupil must be inferred from the responses to the five questions. Pupils assigned to the same class resemble each other in the sense that the probabilities of correctly responding to the five items are the same for these pupils. Between classes these so-called class-specific probabilities are different.

Table 7.8 displays the class-specific probabilities in case there are two latent classes. The probability that a member of class $d = 1, 2$ gives the correct response to question $j = 1, \ldots, 5$ is denoted by π_{dj}. The probability that a pupil is a member of class d is denoted by ω_d. Because the probability of responding correctly to the questions depends on a pupil's arithmetic ability, with two latent classes an obvious expectation is that one of the classes contains pupils with a relatively low arithmetic ability, and the other class contains pupils with a relatively high arithmetic ability. This expectation can be formalized in an informative hypothesis:

$$H_1 : \pi_{1j} < \pi_{2j} \text{ for } j = 1, \ldots, 5, \tag{7.11}$$

that is, in the first class the probabilities of responding correctly are smaller than in the second class. The result is a model with ordered latent classes (Croon, 1990, 1991) that can be seen as the latent class counterpart of Mokken's model for monotone homogeneity (Sijtsma and Molenaar, 2002, Chapters 2 and 3). Looking at the questions, it appears that the first question is easiest, the last question most difficult, and the others ordered in between. This too can be formalized in an informative hypothesis:

$$H_2 : \pi_{d1} > \pi_{d2} > \pi_{d3} > \pi_{d4} > \pi_{d5} \text{ for } d = 1, 2, \tag{7.12}$$

that is, the probability of responding correctly is, in both classes, highest for the first and smallest for the last question. A logical third hypothesis is the combination of both

TABLE 7.8: Parameters of the Two Class Model

Class 1	Class 2
π_{11}	π_{21}
π_{12}	π_{22}
π_{13}	π_{23}
π_{14}	π_{24}
π_{15}	π_{25}
ω_1	ω_2

hypotheses:

$$H_3 : H_1 \& H_2. \tag{7.13}$$

The result is the latent class counterpart of Mokken's model for double monotonicity (Sijtsma en Molenaar, 2002, Chapters 2 and 6).

7.4.2 Specific Features

The density of the data and the prior distributions used for latent class analysis are given in Appendix 10A and Appendix 10B, respectively. A convenient choice is to use noninformative Beta and Dirichlet prior distributions for the class weights and class-specific probabilities, respectively. These are obtained using $a_0 = b_0 = 1$ for both priors (see Appendix 10C).

As for all models considered in this book,

$$BF_{ma} = \frac{f_m}{c_m}. \tag{7.14}$$

However, due to a phenomenon known as label switching (Stephens, 2000), the definition of f_m and c_m must be reconsidered. Consider a hypothetical sample obtained from the posterior distribution displayed in Table 7.9. As can be seen, in the first samples, $\pi_{11} < \pi_{21}$ and $\pi_{12} < \pi_{22}$, which is in accordance with H_1 and states that the probabilities in the first class should be smaller than the probabilities in the second class. In later samples it can be observed that $\pi_{11} > \pi_{21}$ and $\pi_{12} > \pi_{22}$, that is, apparently not in accordance with H_1. The phenomenon illustrated in Table 7.9 is known as label switching. Latent class models are nonidentified in the sense that each permutation of the class labels $d = 1, 2$ renders an equivalent representation. Stated otherwise, to compute f_m and c_m for each vector of class-specific probabilities sampled from the posterior and prior distribution, respectively, one should determine whether there is a permutation that is in agreement with the hypothesis

TABLE 7.9: Hypothetical Sample from the Posterior Distribution

Sample	π_{11}	π_{12}	...	π_{21}	π_{22}	...
1	.34	.21	...	.45	.40	...
2	.32	.25	...	.43	.32	...
3	.29	.21	...	.49	.33	...
4	.30	.23	...	.44	.34	...
.	.	.	.	.	.	.
114	.47	.31	...	.34	.21	...
115	.44	.34	...	.29	.25	...
116	.45	.35	...	.30	.23	...
117	.45	.32	...	.33	.22	...
.	.	.	.	.	.	.

under investigation. The immediate implication is that all parameter vectors displayed in Table 7.9 are in agreement with H_1. This leads to the following modified definitions of f_m and c_m:

Definition 7.1: f_m is the proportion of the posterior distribution of $\pi_{11} \ldots, \pi_{25}$ for which at least one permutation of the class labels renders a set $\pi_{11} \ldots, \pi_{25}$ that is in agreement with H_m.

Definition 7.2: c_m is the proportion of the prior distribution of $\pi_{11} \ldots, \pi_{25}$ for which at least one permutation of the class labels renders a set $\pi_{11} \ldots, \pi_{25}$ that is in agreement with H_m.

7.4.3 Hypothesis Evaluation Using WinBUGS

Appendices 7D and 7E elaborate how the Bayes factors of interest can be computed using WinBUGS. As is elaborated in Section 7.5.3, complexity and fit can be estimated using a sample from the prior and posterior distribution, respectively. Using a sample of $T = 10,000,000$ from the prior distribution using the code in Appendix 7D rendered $c_1 = .062$ and $c_2 = .00007$. Execution of the code Appendix 7E rendered $f_1 = .997$ and $f_2 = .00007$ based on a sample of $T = 10,000,000$ from the posterior distribution. The corresponding Bayes factors are $BF_{1a} = 16.08$ and $BF_{2a} = 1$. Stated otherwise, the assumption of monotonicity is supported by the data, but the assumption of double monotonicity is not. From these results it can be concluded that H_3 is not supported by the data because H_2, one of the components of H_3, is not supported by the data. See also Section 6.5 in in which the evaluation of informative components is elaborated.

Because the data were generated in accordance with Mokken's model of monotone homogeneity, these results are quite satisfactory. Note that the responses of the 40 persons in the top panel of Table 7.7 were generated using $\pi_{11} = \ldots = \pi_{15} = .8$ and of the 40 persons in the bottom panel using $\pi_{21} = \ldots = \pi_{25} = .2$.

As elaborated in Section 7.5.3, the precision with which c_m and f_m (and thus BF_{1a} and BF_{2a}) are estimated depends on the size T of the sample from the prior and posterior distribution. As is elaborated in Appendix 7F, WinBUGS can be used to compute 95% central credibility intervals for BF_{1a} and BF_{2a}, reflecting the uncertainty due to sampling (see Section 10.5.4 for further information). This results in the interval [16.04–16.12] for BF_{1a} and the interval [.77–1.26] for BF_{2a}. As can be seen, using $T = 10,000,000$ renders estimates of both Bayes factors that are very accurate.

7.4.4 Further Reading

The evaluation of informative hypotheses in the context of latent class analysis is discussed in Hoijtink (2001); Laudy, Boom, and Hoijtink (2004); and Hoijtink and Boom (2008). Applications can be found in Laudy, Zoccolillo, Baillargeon, Boom, Tremblay, and Hoijtink (2005) and Van de Schoot and Wong (In Press). Note that in none of these papers is the approach proposed in this book used to compute Bayes factors. This results in a serious drawback that is elaborated in Hoijtink and Boom (2008). Note that hypotheses evaluation using WinBUGS as described in this section does not suffer from this drawback.

7.5   A General Framework

In this book, sofar the evaluation of informative hypotheses using the Bayes factor has been discussed in specific contexts like ANOVA, multiple regression, repeated measures analysis, contingency table analysis, multilevel modeling, and latent class analysis. However, evaluation of informative hypotheses using the Bayes factor is not limited to these contexts. This section contains a nontechnical description of a general framework for the evaluation of informative hypotheses. In Chapter 10 the statistical foundations of Bayesian evaluation of informative hypotheses are elaborated for the univariate and multivariate normal linear models and the general framework.

7.5.1   Data, Model, Parameters

The first step in the evaluation of informative hypotheses using the Bayes factor is the choice of a statistical model for the data $\boldsymbol{Y}, \boldsymbol{X}$ that have been or will be collected. Note that in the context of the univariate normal linear model, $\boldsymbol{Y}$ contains the responses of each person to the dependent variable and $\boldsymbol{X}$ contains the responses of each person to the predictors. Note also that in the context of contingency tables $\boldsymbol{Y}$ denotes the frequencies in each cell of the contingency table and $\boldsymbol{X}$ is empty.

Let $\boldsymbol{\theta}$ denote a vector containing the structural parameters of the chosen model, that is, the parameters that will be used in the formulation of informative hypotheses, and let $\boldsymbol{\phi}$ denote the nuisance parameters, that is, the model parameters that will not be used in the formulation of informative hypotheses. Note that in the univariate normal linear model, $\boldsymbol{\theta}$ contains the regression coefficients and $\boldsymbol{\phi}$ the residual variance. Note also that for contingency tables $\boldsymbol{\theta}$ contains the cell probabilities and $\boldsymbol{\phi}$ is empty. The density of the data can be written as

$$f(\boldsymbol{Y} \mid \boldsymbol{X}, \boldsymbol{\theta}, \boldsymbol{\phi}). \tag{7.15}$$

Examples of densities of the data can be found in Appendix 10A.

An important consideration when formulating the density of the data is the following:

Consideration 7.1: Make sure that the parameters in $\boldsymbol{\theta}$ are comparable, that is, that it is meaningful to constrain them using equality and inequality constraints.

As elaborated in Section 1.4.1, unstandardized regression coefficients are an example of noncomparable parameters.

7.5.2   Informative Hypotheses and the Prior Distribution

Once the density of the data and $\boldsymbol{\theta}$ have been specified, informative hypotheses can be formulated:

$$H_m : \boldsymbol{R}_m t(\boldsymbol{\theta}) > \boldsymbol{r}_m \text{ and } \boldsymbol{S}_m t(\boldsymbol{\theta}) = \boldsymbol{s}_m, \tag{7.16}$$

where $t(\cdot)$ denotes a monotone transformation of each element of the vector $\boldsymbol{\theta}$, which contains D parameters, or a monotone transformation of groups of elements of the same size; $\boldsymbol{R}_m$ and $\boldsymbol{S}_m$ are $Q \times D$ and $L \times D$ matrices, respectively, where Q denotes the number of inequality constraints and L the number of equality constraints; and $\boldsymbol{r}_m$ and $\boldsymbol{s}_m$ are vectors of length Q and L, respectively.

Note that in the context of contingency tables, using $t(\pi_d) = \log \pi_d = \theta_d$, the constraint

$$\frac{\pi_1 \pi_2}{\pi_3 \pi_4} > 1 \tag{7.17}$$

can be reformulated as

$$\theta_1 + \theta_2 - \theta_3 - \theta_4 > 0, \tag{7.18}$$

which is of the form (7.16). Note also that constraints of the form $H_m : \boldsymbol{R}_m \boldsymbol{\theta} > \boldsymbol{r}_m$ and $\boldsymbol{S}_m \boldsymbol{\theta} = \boldsymbol{s}_m$, that is, without a transformation $t(\cdot)$ of $\boldsymbol{\theta}$, were considered in Chapters 1 through 6.

If informative hypotheses are specified only using inequality constraints, then

- If possible, a noninformative normal prior distribution should be used for each element of $\boldsymbol{\theta}$, that is, $h(\theta_d \mid H_a) \sim \mathcal{N}(\mu_0, \tau_0^2)$ for $d = 1, \ldots, D$, with $\mu_0 = 0$ and $\tau_0^2 \to \infty$. A standard noninformative prior distribution should be used for the nuisance parameters ϕ.

- If a normal prior distribution is not appropriate for the parameters at hand, use a uniform distribution or a distribution approaching uniformity for each parameter under consideration. For example, for each of the class-specific probabilities in a latent class model, a uniform prior on the interval $[0, 1]$ can be used. A standard noninformative prior distribution should be used for the nuisance parameters ϕ.

If the specification of informative hypotheses includes equality constraints, then

- If the parameters are naturally bounded, like, for example, probabilities, use a uniform distribution over their domain. A standard noninformative prior distribution should be used for the nuisance parameters ϕ.

- If the parameters are not naturally bounded, as was the case for all examples discussed in Chapters 1 through 6, researchers can either use one of the statistical packages presented in Chapter 9 (but not `WinBUGS`) or consult an experienced statistician.

7.5.3 Hypotheses Evaluation Using `WinBUGS`

If informative hypotheses are formulated only using inequality constraints, `WinBUGS` (see Section 9.2.7) can be used to compute Bayes factors. The previous two sections of this chapter contained illustrations of the computation of Bayes factors using `WinBUGS`. Three steps must be executed:

- Obtain a sample of size T from the unconstrained prior and posterior distribution.

- Use this sample to determine c_m and f_m for each informative hypothesis under consideration, that is, determine the proportion of the sample from the unconstrained prior and posterior distribution in agreement with each of the hypotheses under consideration. If models are inherently unidentified, like the latent class model discussed in the previous section, the following consideration must be accounted for:

Consideration 7.2: Determine for the statistical model at hand whether it is identified or not, that is, are there statistically equivalent representations of each parameter vector. If the answer is yes, make sure to account for these phenomena when determining whether a sampled parameter vector is in agreement with the hypotheses under investigation or not.

TABLE 7.10: D versus T

D	2	3	4	5	6	7	8
True c_m	1/2	1/6	1/24	1/120	1/720	1/5040	1/40320
	=.5	=.166	=.042	= .008	=.0014	= .0002	=.000025
T	1,000	3,000	9,600	120,000	360,000	2,520,000	20,160,000
LB-95%	.47	.154	.038	.0078	.00127	.000182	.000023
UB-95%	.53	.180	.046	.0089	.00152	.000217	.000027

- Compute the required Bayes factors using one the following formulae: for the comparison of an informative hypothesis H_m with the unconstrained alternative H_a, use $BF_{ma} = f_m/c_m$; for the comparison of an informative hypothesis with its complement H_{m_c}, use $BF_{mm_c} = (f_m/c_m)/((1-f_m)/(1-c_m))$; and for the comparison of an informative hypothesis with a competing informative hypothesis $H_{m'}$, use $BF_{mm'} = BF_{ma}/BF_{m'a}$.

This procedure will work fine if hypotheses are only specified using inequality constraints (and not using equality, about equality, or range constraints) if the number of parameters D and the number of constraints Q are relatively small. For larger values of D and Q, the size of T, needed to obtain accurate estimates of c_m and f_m, becomes so large that the procedure may no longer be of practical value. The question how large D and Q may be such that simply sampling from the prior and posterior of the unconstrained model will render accurate estimates of c_m and f_m is now considered. First, estimation of c_m will be discussed, subsequently the estimation of f_m.

The larger Q, the smaller c_m and the larger the sample that is needed in order to be able to accurately estimate c_m. Therefore the elaboration below is based on a complete ordering of the D parameters, that is, $Q = D - 1$ with $H_m : \theta_1 > \ldots > \theta_D$. If T is large enough to accurately estimate c_m for this hypothesis, it will also be large enough to estimate c_m for informative hypotheses based on fewer inequality constraints. In Table 7.10 the sizes T of the sample from the prior distribution needed in order to accurately estimate c_m for $D = 2, \ldots, 8$ are displayed. Note that, by accurately it is meant that the difference between the true c_m and the lower bound and upper bound of a 95% central credibility interval for the estimate of c_m is less than 10%. Assuming that f_m is exactly known, this implies that $BF_{ma} = f_m/c_m$ is never off by more than 10%.

The following can be concluded from Table 7.10:

- The complexity c_m can accurately be evaluated for a complete ordering of six parameters using a sample of $T = 360,000$ from the prior distribution. The implication is that c_m can almost exactly be computed for any inequality constrained hypothesis based on $D = 6$ parameters using a sample of $T = 1,000,000$ from the prior distribution. Note that samples of $T = 1,000,000$ and their evaluation is no problem if WinBUGS is used.

- Accurate computation of complexity for a complete ordering of seven parameters can be achieved using WinBUGS (enough patience will render a sample of, say, 5,000,000 and its evaluation). For more than seven parameters, accurate computation of c_m for a complete ordering may not always be possible.

- Note that not always a complete ordering of parameters is of interest. Consider, for example, $\theta_1 > \theta_2 > \theta_3 > \theta_4, \theta_5 > \theta_6 > \theta_7 > \theta_8$, the combination of two independent sets of constraints. The complexity of this hypothesis is $1/24 \times 1/24 = .0017$. According to Table 7.10, a complexity of about .0017 can accurately be estimated using $T = 360,000$ and virtually exactly using $T = 1,000,000$.

- As exemplified in the previous bullet, Table 7.10 does not only apply to hypotheses in which the parameters at hand are completely ordered. As long as there is an idea of the size of the complexity to be estimated, Table 7.10 gives an indication of the sample sizes needed in order to obtain an accurate estimate.

Now the T rendering accurate estimates of c_m has been discussed, the implications for f_m can be elaborated:

- If the sample from the prior is of size T, use also a sample from the posterior of size T.

- If the fit of a hypothesis is really bad, irrespective of the size of T, the number of parameter vectors sampled from the posterior distribution that are in agreement with H_m will be very small. This is not a problem if one only wants to compare the hypothesis at hand to the unconstrained alternative using BF_{ma} or the complementary hypothesis using BF_{mm_c}. The conclusion will simply be that H_m is not supported by the data. However, one can no longer use $BF_{mm'}$ to determine which of two badly fitting competing hypotheses is better. The estimates of f_m and f'_m will be so unstable that a comparison of both hypotheses is no longer possible.

Finally, note that in the previous two sections and Appendices 7C and 7F it was exemplified how WinBUGS can be used to compute a credibility interval around each Bayes factor in order to quantify its uncertainty due to Monte Carlo sampling (see Section 10.5.4 for further information). Application of this procedure can be used as a final check to verify whether T is large enough for the hypotheses at hand.

7.5.4 Further Reading

The statistical foundation of the evaluation of informative hypotheses both for univariate and multivariate normal linear models and for the general framework will be presented in Chapter 10. In Section 9.2.7, the use of WinBUGS for the evaluation of informative hypotheses is elaborated (see also Van Rossum, van de Schoot, and Hoijtink, In Press).

7.6   Appendices: Sampling Using Winbugs

7.6.1   Appendix 7A: Prior Sampling for the Multilevel Model

This appendix provides annotated WinBUGS code for sampling from the prior distribution of a multilevel model for four repeated measurements describing weight development over time.

```
model {

# Specification of a multivariate normal prior distribution for
# the means of the random intercepts and slopes. See Appendix 10B.
# Note that WinBUGS specifies the multivariate
# normal distribution using a precision matrix IT0 that is
# the inverse of T0.

mu[1:2] ~ dmnorm(mu0[],IT0[ , ])

# Determine whether or not a vector mu sampled from the prior
# distribution is in agreement with H1. If the sampled mu
# is in agreement with the first element of H1, c1=1 and 0
# otherwise. If the sampled mu is in agreement with the second
# element of H1 c2=1 and 0 otherwise. WinBUGS will give the
# proportion of complex equal to 1, that is, the complexity
# of H1. To obtain the complexity of H2 change the second
# line below to: c2 <- step(10-3 *mu[2]).

c1 <- step(mu[1]-100)
c2 <- step(3 *mu[2]-10)
complex <- c1 * c2

}

# Specification of the parameters of the prior distribution,
# that is mu0 and IT0. Note that, the precision matrix IT0
# is the inverse of the covariance matrix T0, and
# consequently that the diagonal elements should approach 0
# instead of infinity.

list(mu0 = c(0,0),
IT0 = structure(.Data = c(1.0E-6, 0, 0, 1.0E-6),
.Dim = c(2, 2)))
```

7.6.2   Appendix 7B: Posterior Sampling for the Multilevel Model

This appendix provides annotated `WinBUGS` code for sampling from the posterior distribution of a multilevel model for four repeated measurements describing weight development over time.

```
model {

# Below the density of the data of the multilevel model is
# specified. See also Appendix 10A

for( i in 1 : N ) {beta[i , 1:2] ~ dmnorm(mu[],Iphi[ , ])
for( t in 1 : M ) y[i , t] ~ dnorm(pred[i , t],Isigma2)
pred[i , t] <- beta[i,1] + beta[i,2] * time[t] }
}


# Below the prior distributions of mu, Iphi and Isigma2 are
# specified, that is, a multivariate normal (the same as in
# Appendix 7A), Wishart and gamma distribution, respectively.
# Note that Iphi is a precision matrix that is, the inverse
# of the covariance matrix phi. Therefore a Wishart instead of
# an inverse Wishart prior distribution is used. Due to the
# small degrees of freedom a vague prior distribution is
# obtained. Note also that Isigma2 is a precision, that is the
# inverse of the variance sigma2. Therefore a gamma instead of
# a scaled inverse chi-square distribution is used. The hyper
# parameters are chosen such that a vague prior distribution
# is obtained. See also Appendix 10B.

mu[1:2] ~ dmnorm(mu0[],IT0[ , ])
Iphi[1:2 , 1:2] ~ dwish(S0[ , ], 2)
Isigma2 ~ dgamma(.01,.01)

# Below the fit of H_1 is computed.

f1 <- step(mu[1]-100)
f2 <- step(3 * mu[2]-10)
fit <- f1 * f2


}

# Specification of subsequently the time-points,
# the sample sizes, the prior parameters and the data

list(time = c(-1.5, -.5, .5, 1.5),
N = 12,
M = 4,
S0 = structure(.Data = c(723.55,-17.06,-17.06,5.12), .Dim = c(2, 2)),
mu0 = c(0,0),
IT0 = structure(.Data = c(1.0E-6, 0, 0, 1.0E-6),
.Dim = c(2, 2)),
```

```
y = structure(.Data = c(
101 , 105 , 110 , 115 ,
140 , 142 , 141 , 143 ,
132 , 140 , 141 , 140 ,
125 , 128 , 133 , 138 ,
160 , 158 , 165 , 166 ,
179 , 178 , 173 , 174 ,
147 , 152 , 156 , 160 ,
143 , 150 , 155 , 155 ,
110 , 112 , 111 , 112 ,
115 , 117 , 121 , 119 ,
120 , 119 , 126 , 122 ,
126 , 130 , 133 , 130),
.Dim = c(12,4)))
```

7.6.3 Appendix 7C: Central Credibility Interval for Bayes Factors

This appendix provides annotated WinBUGS code for the computation of a 95% central credibility interval for BF_{12}.

```
model {

# Below enter as parameters for the beta distribution
# cm * T + 1 and T - cm * T + 1
c1 ~ dbeta(229001,771001)

# Below enter as parameters for the beta distribution
# fm * T + 1 and T - fm * T + 1
f1 ~ dbeta(44001,956001)


# Use WinBUGS to compute the 2.5th and 97.5th percentile of bf1.

bf1<- f1/c1

# Below enter as parameters for the beta distribution
# cm * T + 1 and T - cm * T + 1
c2 ~ dbeta(231001,769001)

# Below enter as parameters for the beta distribution
# fm * T +1 and T - fm * T + 1
f2 ~ dbeta(950001,50001)

# Use WinBUGS to compute the 2.5th and 97.5th percentile of bf2.

bf2 <- f2/c2

# Use WinBUGS to compute the 2.5th and 97.5th percentile of bf12.

bf12 <- bf1/bf2

}
```

7.6.4   Appendix 7D: Prior Sampling for the Latent Class Model

This appendix provides annotated WinBUGS code for sampling from the prior distribution of a latent class model with $D = 2$ classes and $J = 5$ questions.

```
model {

# Below a uniform prior distribution on the interval [0,1]
# is assigned to each class-specific probability. See also
# Appendix 10B, where a uniform distribution is
# obtained using a_0 = b_0 = 1 for the Beta distribution.

p[1,1] ~ dunif(0,1)
p[1,2] ~ dunif(0,1)
p[1,3] ~ dunif(0,1)
p[1,4] ~ dunif(0,1)
p[1,5] ~ dunif(0,1)
p[2,1] ~ dunif(0,1)
p[2,2] ~ dunif(0,1)
p[2,3] ~ dunif(0,1)
p[2,4] ~ dunif(0,1)
p[2,5] ~ dunif(0,1)

# Below it is verified for the first permutation of the class
# labels whether a sample from the prior is in agreement with H_1.
# If the answer is yes, c1=c2=c3=c4=c5=1 and thus complex1=1,
# otherwise one of the c-s and thus complex1 is equal to 0.
# Note that WinBUGS gives the proportion of complex1=1
# that is, the complexity of H_1, in its results.

c1 <- step(p[1,1]-p[2,1])
c2 <- step(p[1,2]-p[2,2])
c3 <- step(p[1,3]-p[2,3])
c4 <- step(p[1,4]-p[2,4])
c5 <- step(p[1,5]-p[2,5])

complex1 <- c1*c2*c3*c4*c5

# Below it is verified for the second permutation of the
# class labels whether a sample from the prior is in agreement
# with H_1. Note that the complexity of H_1 if the sum of
# the proportions complex1 and complex2 in agreement with H_1

c6 <- step(p[2,1]-p[1,1])
c7 <- step(p[2,2]-p[1,2])
c8 <- step(p[2,3]-p[1,3])
c9 <- step(p[2,4]-p[1,4])
c10 <- step(p[2,5]-p[1,5])

complex2 <- c6*c7*c8*c9*c10
```

```
# Below it is verified whether a sample from the prior
# is in agreement with H2.

c11 <- step(p[1,1]-p[1,2])
c12 <- step(p[1,2]-p[1,3])
c13 <- step(p[1,3]-p[1,4])
c14 <- step(p[1,4]-p[1,5])
c15 <- step(p[2,1]-p[2,2])
c16 <- step(p[2,2]-p[2,3])
c17 <- step(p[2,3]-p[2,4])
c18 <- step(p[2,4]-p[2,5])

complex3 <-c11*c12*c13*c14*c15*c16*c17*c18

# Below it is verified whether a sample from the prior
# is in agreement with H3. The complexity is the sum of
# the proportions of complex4=1 and complex5=1.

complex4 <- complex1 * complex3

complex5 <- complex2 * complex3

}
```

7.6.5   Appendix 7E: Posterior Sampling for the Latent Class Model

This appendix provides annotated WinBUGS code for sampling from the posterior distribution of a latent class model with $D = 2$ classes and $J = 5$ questions.

```
model {

# Below the density of the data of the latent class model is
# specified. See also Appendix 10A. Note that ksi[i]
# denotes the latent class membership of person i.

for( i in 1 : N ) {ksi[i] ~ dcat(Omega[])

for (j in 1:M) {a[i,j] ~ dbern(p[ksi[i],j])}}

# Below a uniform prior distribution on the interval [0,1]
# is assigned to each class-specific probability and a
# Dirichlet prior for the class weights. See, Appendix 10B.

Omega[1:2] ~ ddirch(alpha[])
p[1,1] ~ dunif(0,1)
p[1,2] ~ dunif(0,1)
p[1,3] ~ dunif(0,1)
p[1,4] ~ dunif(0,1)
p[1,5] ~ dunif(0,1)
p[2,1] ~ dunif(0,1)
p[2,2] ~ dunif(0,1)
p[2,3] ~ dunif(0,1)
p[2,4] ~ dunif(0,1)
p[2,5] ~ dunif(0,1)

# Below it is verified whether a sample from the posterior
# is in agreement with H_1 for the first permutation of
# the class labels.

f1 <- step(p[1,1]-p[2,1])
f2 <- step(p[1,2]-p[2,2])
f3 <- step(p[1,3]-p[2,3])
f4 <- step(p[1,4]-p[2,4])
f5 <- step(p[1,5]-p[2,5])

fit1 <- f1*f2*f3*f4*f5

# Below it is verified whether a sample from the posterior
# is in agreement with H_1 for the second permutation
# of the class labels.

f6 <- step(p[2,1]-p[1,1])
f7 <- step(p[2,2]-p[1,2])
f8 <- step(p[2,3]-p[1,3])
f9 <- step(p[2,4]-p[1,4])
```

```
f10 <- step(p[2,5]-p[1,5])

fit2 <- f6*f7*f8*f9*f10

# Below it is verified whether a sample from the posterior
# is in agreement with H2.

f11 <- step(p[1,1]-p[1,2])
f12 <- step(p[1,2]-p[1,3])
f13 <- step(p[1,3]-p[1,4])
f14 <- step(p[1,4]-p[1,5])
f15 <- step(p[2,1]-p[2,2])
f16 <- step(p[2,2]-p[2,3])
f17 <- step(p[2,3]-p[2,4])
f18 <- step(p[2,4]-p[2,5])

fit3 <-f11*f12*f13*f14*f15*f16*f17*f18

# Below it is verified whether a sample from the posterior
# is in agreement with H3.

fit4 <- fit1 * fit3

fit5 <- fit2 * fit3
}

# Below the data are specified for 80 cases and 5 questions.
# Furthermore the hyper parameters alpha of the Dirichlet prior
# are specified such that a vague distribution is obtained.

list(N = 80,M=5,alpha=c(1,1),a=structure(.Data=c(
1,1,1,0,1,
1,0,1,1,1,
1,1,1,1,1,
...
0,0,1,0,0,
0,1,0,1,1,
0,0,0,0,1),
.Dim=c(80,5)))
```

7.6.6   Appendix 7F: Credibility Interval for Bayes Factors

This appendix provides annotated WinBUGS code for the computation of a 95% central credibility interval for BF_{1a} and BF_{2a}.

```
model {

# Below enter as parameters for the beta distribution
# cm * T + 1 and T - cm * T + 1.

c1 ~ dbeta(620001,9380001)

# Below enter as parameters for the beta distribution
# fm * T + 1 and T - fm * T + 1.

f1 ~ dbeta(997001,3001)

# Use WinBUGS to compute the 2.5th and 97.5th percentile of bf1a.

bf1a<- f1/c1

# Below enter as parameters for the beta distribution
# cm * T + 1 and T - cm * T + 1.

c2 ~ dbeta(701,9999301)

# Below enter as parameters for the beta distribution
# fm * T + 1 and T - fm * T + 1.

f2 ~ dbeta(71,999931)

# Use WinBUGS to compute the 2.5th and 97.5th percentile of bf2a.

bf2a <- f2/c2
}
```

Chapter 8

Other Approaches

8.1   Introduction

Part II and Chapter 7 of this book dealt with Bayesian evaluation of informative hypotheses. In this chapter two other approaches for the evaluation of informative hypotheses are introduced. The first approach evaluates informative hypotheses by means of p-values, that is, null hypothesis significance testing. An important reference for this approach is Silvapulle and Sen (2004). The second approach provides an evaluation by means of a modification of Akaike's information criterion (Akaike, 1987) such that it can be applied to the evaluation of informative hypotheses. Important references for this approach are Anraku (1999); Kuiper, Hoijtink, and Silvapulle (In Press); Kuiper, Hoijtink, and Silvapulle (Unpublished); and Kuiper and Hoijtink (Unpublished).

A nontechnical introduction of each approach is provided using an example originating in Lucas (2003) previously used to illustrate Kuiper, Klugkist, and Hoijtink (2010). Lucas (2003) describes an experiment in which the influence a leader has on the members in his group is scored by each group member on a five-point Likert scale. 150 persons were randomly divided over five groups. In the first group, a man was randomly chosen and appointed leader. In the second group, a woman was randomly appointed. In the third and fourth groups, a man and woman, respectively, were chosen based on skills proven to the group. In the fifth group, a skilled woman was appointed after female leadership was institutionalized, that is, after the group was primed by showing them examples of successful female leaders. The main research question of Lucas (2003) was whether institutionalization of female leadership reduces the influence gap between men and women. The following hypotheses can be derived from his paper (note that μ_1 denotes the mean influence level in the first group):

$$H_1 : \{\mu_1, \mu_3\} > \{\mu_2, \mu_4, \mu_5\}, \qquad (8.1)$$

that is, the influence of men is always larger than the influence of women;

$$H_2 : \mu_3 = \mu_5 > \{\mu_1, \mu_4\} > \mu_2, \qquad (8.2)$$

TABLE 8.1: Descriptive Statistics for the Lucas (2003) Data

Group	Mean	Standard Deviation	N
1 Random Man	2.33	1.86	30
2 Random Woman	1.33	1.15	30
3 Able Man	3.20	1.79	30
4 Able Woman	2.23	1.45	30
5 Able Institutionalized Woman	3.23	1.50	30

that is, institutionalization works and men and skilled women have more influence than randomly appointed women. Together with the traditional null and alternative hypothesis

$$H_0 : \mu_1 = \mu_2 = \mu_3 = \mu_4 = \mu_5, \tag{8.3}$$

$$H_a : \mu_1, \mu_2, \mu_3, \mu_4, \mu_5, \tag{8.4}$$

there are four hypotheses that will be investigated. Descriptive statistics for the data from Lucas (2003) can be found in Table 8.1.

All analyses presented in this chapter are executed using `ConfirmatoryANOVA` (see Kuiper, Klugkist and Hoijtink (2010) and Section 9.2.1). In Appendix 8A the data and command file needed to run `ConfirmatoryANOVA` are presented. Note that the command file can be constructed manually or using a `Windows` user interface. The results obtained using the Bayesian approach, null hypothesis significance testing, and the order-restricted information criterion are presented in Table 8.2. The main features of each approach are presented in Table 8.3, which will be the point of departure for discussion, evaluation, and comparison of the Bayesian approach, null hypothesis significance testing, and information criteria.

8.2   Resume: Bayesian Evaluation of Informative Hypotheses

Bayesian evaluation of informative hypotheses was extensively discussed in Part II and Chapter 7 of this book. Here straightforwardly the results of the evaluation of the four hypotheses of interest in this chapter and the main features of the Bayesian approach are presented.

As can be seen in Table 8.2 in the columns labeled Bayes Factor (BF) and Posterior Model Probability (PMP), the support in the data for H_2 is 79.89 times larger than the support for H_a, and $79.89/.13 = 614.54$ times larger than for H_1. This constitutes

TABLE 8.2: Evaluation of H_0, H_1, H_2 and H_a Using Three Approaches Implemented in `ConfirmatoryANOVA`

Hypothesis	BF vs H_a	PMP	p-value vs H_0	p-value vs H_a	ORIC
H_0	.00	.00			588.54
H_1	.13	.00	.00	.09	569.50
H_2	79.89	.99	.00	1.00	562.50
H_a	1.00	.01			568.10

a lot of evidence in favor of H_2, which is also expressed by the PMP of .99. Note that `ConfirmatoryANOVA` uses a prior distribution of the form

$$h(\mu_1, \mu_2, \mu_3, \mu_4 \mid H_a)h(\sigma^2) = \mathcal{N}(\mu_0, \tau_0^2)\mathcal{N}(\mu_0, \tau_0^2)\mathcal{N}(\mu_0, \tau_0^2)\mathcal{N}(\mu_0, \tau_0^2)\text{Inv-}\chi^2(\nu_0, \sigma_0^2). \quad (8.5)$$

For the Lucas data, the specification is $\mu_0 = 2.28$, $\tau_0^2 = 2.32$, $\nu_0 = 1$, and $\sigma_0^2 = 2.5$.

The main features of the Bayesian approach are summarized in Table 8.3. However, before discussing these features, the entries of Table 8.3 are introduced. The table presents the properties of each of five methods for the evaluation of informative hypotheses. Note that "Y" denotes that a property applies to the method at hand, and that "?" denotes that the property is further discussed in the section addressing the method at hand. The top panel addresses the constraints that can be used for the construction of informative hypotheses:

- Simple Inequality, for example, $\mu_1 > \mu_2$, that is, inequality constraints between pairs of means.

- Simple Equality, for example, $\mu_1 = \mu_2$, that is, equality constraints between pairs of means.

- About Equality, for example, $|\mu_1 - \mu_2| < d \times \sigma$, where d denotes Cohen's effect size for the difference between two means and σ the within group standard deviation.

- Range Constraints, for example, $\mu_1 - \mu_2 > .2 \times \sigma$ combined with $\mu_1 - \mu_2 < .5 \times \sigma$, that is, μ_1 is between a .2 and .5 standard deviations larger than μ_2.

- Effect Sizes, for example, $\mu_1 > \mu_2 + d \times \sigma$, that is, μ_1 is at least d standard deviations larger than μ_2.

- Combinations, for example, $\mu_1 - \mu_2 > \mu_3 - \mu_4$, that is, the difference between the first two means is larger than the difference between the last two means.

The middle panel addresses the hypotheses to which an informative hypothesis can be compared using the method at hand:

- H_0, that is, the null hypothesis; for the example at hand, see (8.3).

- H_a, that is, the unconstrained alternative hypothesis; for the example at hand, see (8.4).

- $H_{m'}$, that is, a competing informative hypothesis; for the example at hand, H_1 is a competing informative hypothesis for H_2.

- H_{m_c}, that is, the complement of an informative hypothesis; for (8.2), the complement is H_{2_c} : not H_2.

The bottom panel addresses other properties of the five methods:

- Prior Distribution, that is, is it necessary to specify a prior distribution?

- $M > 2$, that is, is a simultaneous evaluation of H_0, H_1, H_2, and H_a possible?

- Evidence, that is, is the support in the data for each hypothesis quantified?

TABLE 8.3: Properties of Methods for the Evaluation of Informative Hypotheses

Method	*F*-bar	Chi-bar	ORIC	GORIC	Bayes Factor
Constraints Available for the Construction of H_m					
Simple Inequality	Y	Y	Y	Y	Y
Simple Equality	?	?	Y	Y	Y
About Equality					Y
Range Constraints					Y
Effect Sizes		?		Y	Y
Combinations		Y		Y	Y
Hypotheses to which H_m can be Compared					
H_0	Y	Y	Y	Y	Y
H_a	Y		Y	Y	Y
$H_{m'}$			Y	Y	Y
H_{m_c}					Y
Other Properties					
Prior Distribution					Y
$M > 2$			Y	Y	Y
Evidence					Y
Statistical Models	ANOVA	GLM/GMM	ANOVA	UNLM MNLM	UNLM/MNLM CONTIN LCA/GLM/GMM
Software	CONFA	SAS PROC PLM	CONFA	GORIC	CONFA/BIEMS/ CONTIN/WinBUGS

- Statistical Models, that is, for which statistical models is software with the method at hand implemented available? The abbreviations used are ANOVA for analysis of variance (see Chapter 1); GLM for generalized linear models (McCullagh and Nelder, 1989); GMM for generalized mixed models (McCullogh and Searle, 2001); UNLM/MNLM for univariate/multivariate normal linear model (see Chapters 1/2); CONTIN for contingency table analysis (see Chapter 7); and LCA for latent class analysis (see Chapter 7).

- Software, that is, the name of the software packages in which the method at hand is implemented. Note that software is further elaborated in the next chapter. Note also that CONFA denotes `ConfirmatoryANOVA` and that CONTIN denotes `ContingencyTable`.

Returning to the main features of the Bayesian approach, it can be seen in the sixth column of Table 8.3 that all constraint types can be used for the specification of an informative hypothesis, that all hypotheses to which H_m can be compared can be used, and that software is available for many statistical models. However, note that a prior distribution must be specified before the Bayesian approach can be used. This implies that prior sensitivity is an issue if hypotheses are formulated using equality, about equality, or range constraints. However, as shown in Chapter 4, prior sensitivity seems to be a minor issue in the context of Bayesian evaluation of informative hypotheses. It also implies that there are hypotheses that cannot be evaluated using the Bayesian approach because the unconstrained prior distributions corresponding to these hypotheses are incompatible (see Section 4.4.2 for an example). A simple example of such hypotheses is $H_0 : \mu = 0$ and $H_1 : \mu > 2$. The prior distribution corresponding to H_0 has a prior mean of 0, and the prior distribution corresponding to H_1 has a prior mean of 2.

8.3 Null Hypothesis Significance Testing

Three books mark the development of and provide a comprehensive overview of the evaluation of informative hypotheses by means of p-values: Barlow, Bartholomew, Bremner, and Brunk (1972); Robertson, Wright, and Dykstra (1988); and Silvapulle and Sen (2004). The interested reader is also referred to Molenberghs and Verbeke (2007) for a short and enlightening introduction to the testing of constrained hypotheses, and to Schervish (1996) for a paper on *"p values what they are and what they are not."* Note that none of these authors called hypotheses formulated using equality and inequality constraints *informative hypotheses*, but each addresses exactly these hypotheses. There is a wealth of interesting ideas, concepts, derivations, and applications in these books. However, here we focus on the evaluation of informative hypotheses in the context of ANOVA models and one generalization. Our focus is motivated by the availability of software for the situations that will be discussed, such that researchers can actually use what is described below for the analysis of their own data. Our focus is not motivated by the statistical theory, which encompasses much more than is currently implemented in software.

8.3.1 ANOVA

Consider analysis of variance as discussed in Section 1.2. With respect to $j = 1, \ldots, J$ means μ_j there are two options to evaluate an informative hypothesis:

$$H_0 : \mu_1 = \ldots = \mu_J \tag{8.6}$$

versus

$$H_m : \boldsymbol{R}_m \boldsymbol{\mu} > 0, \tag{8.7}$$

that is, the informative hypothesis is the alternative hypothesis, and

$$H_m : \boldsymbol{R}_m \boldsymbol{\mu} > 0 \tag{8.8}$$

versus

$$H_a : \mu_1, \ldots, \mu_J, \tag{8.9}$$

that is, the informative hypothesis is the null hypothesis.

Note that $\boldsymbol{R}_m$ is a $Q \times J$ matrix, where Q denotes the number of rows in $\boldsymbol{R}_m$ needed to construct H_m. In the software package `ConfirmatoryANOVA`, each row of $\boldsymbol{R}_m$ must have exactly two non-zero elements, one with the value 1 and one with the value -1. For example, with $J = 3$, the row $1 \; -1 \; 0$ would denote the constraint $1 \times \mu_1 - 1 \times \mu_2 + 0 \times \mu_3 > 0$, that is, $\mu_1 > \mu_2$. Stated otherwise, only simple constraints can be used to specify informative hypotheses. In the next section this requirement with respect to $\boldsymbol{R}_m$ will be relaxed.

Hypothesis significance testing using the F-bar statistic (see Chapter 2 of Silvapulle and Sen (2004)) can be used to evaluate (8.6) with respect to (8.7), the result is a p-value. As elaborated in Section 5.2.1 (see Chapter 2 in Silvapulle and Sen (2004)), the alpha-level is the probability of incorrectly rejecting H_0 if it is true. If the p-value is smaller than the alpha-level chosen, say .05, which has become a golden standard by convention, the null hypothesis is rejected in favor of the alternative. If the p-value is larger than the alpha-level, the null hypothesis is not rejected. If hypothesis testing using the F-bar statistic is applied to the evaluation of (8.8) with respect to (8.9), the result is also a p-value.

The second column of Table 8.3 summarizes the main features of null hypothesis significance testing as implemented in `ConfirmatoryANOVA`: one informative hypothesis specified using simple inequality constraints can be evaluated against H_0 and H_a by means of a p-value in the context of ANOVA models. Note that the question mark in the column labeled F-bar stresses that `ConfirmatoryANOVA` also allows the use of simple equality constraints for the specification of an informative hypothesis.

As can be seen in the fourth column of Table 8.2, application of null hypothesis significance testing to the running example renders H_0 being rejected in favor of both H_1 and H_2. As can be seen in the fifth column, neither H_1 nor H_2 is rejected in favor of H_a (using an alpha-level of .05). The conclusion is clear: both H_1 and H_2 are preferred over H_0 and H_a. However, a mutual comparison of H_1 with H_2 is not possible using null hypothesis significance testing.

8.3.2   A General Class of Models

`SAS PROC PLM` (see Section 9.2.6) enables comparison of

$$H_0 : \boldsymbol{R}_m \boldsymbol{\theta} = 0 \tag{8.10}$$

versus

$$H_m : \boldsymbol{R}_m \boldsymbol{\theta} > 0. \tag{8.11}$$

by means of null hypothesis significance testing based on the chi-bar-squared statistic (Silvapulle and Sen, 2004, Chapter 3.4), if under the null hypothesis

$$\boldsymbol{R}_m \hat{\boldsymbol{\theta}} \sim \mathcal{N}(\boldsymbol{0}, \boldsymbol{\Sigma}). \tag{8.12}$$

Note that $\boldsymbol{\theta}$ denotes the parameters (e.g., means, regression coefficients) of the statistical model (e.g., multiple regression, repeated measures analysis) providing the context for the specification of the informative hypotheses. The requirement formulated in (8.12) states that if data matrices are repeatedly sampled from a population in which (8.10) is true, the distribution of the estimate of $\boldsymbol{R}_m \boldsymbol{\theta}$ for each of these data matrices should be a multivariate normal distribution with mean vector $\boldsymbol{0}$ and covariance matrix $\boldsymbol{\Sigma}$.

It is well-known that estimates of the parameters, that is, the means and regression coefficients, of the multivariate normal linear model are distributed according to a multivariate t-distribution if the null hypothesis is true. For relatively small sample sizes, the multivariate t can be accurately approximated by a multivariate normal distribution. Using the fact that linear combinations of normally distributed random variables like $\boldsymbol{R}_m \boldsymbol{\theta}$ also have a normal distribution, this explains why the multivariate normal linear model is in close agreement with the requirement formulated in (8.12). For many other statistical models, for example, the generalized linear model (McCullagh and Nelder, 1989) and the generalized mixed model (McCullogh and Searle, 2001), it also holds that asymptotically parameter estimates are normally distributed if the null hypothesis is true. However, there is room for research with respect to the size of the sample needed such that the estimates of the parameters of the model at hand are in agreement with the requirement formulated in (8.12).

A situation in which the method implemented in `SAS PROC PLM` can be applied is a univariate regression with two predictors:

$$Z(y_i) = \beta_0 + \beta_1 Z(x_{1i}) + \beta_2 Z(x_{2i}) + \epsilon_i. \tag{8.13}$$

Note that $Z(y_i)$, $Z(x_{1i})$ and $Z(x_{2i})$ denote the standardized scores of the i-th person on the dependent variable and both predictors, respectively. Note also that β_1 and β_2 are standardized regression coefficients relating the predictors to the dependent variable. Note finally that the residuals ϵ_i have a normal distribution with mean zero and variance σ^2. Let

$\boldsymbol{\theta} = \{\beta_0, \beta_1, \beta_2\}$. Using an $\boldsymbol{R}_m$-matrix with the row $[0 \ 1 \ -1]$ the null hypothesis obtained is $H_0 : \beta_1 = \beta_2$ and the informative hypothesis is $H_m : \beta_1 > \beta_2$. As in the previous section, a p-value is used to evaluate H_0 with respect to H_m. Note that H_m is only a meaningful hypothesis if x_1 and x_2 are measured on the same scale. This was achieved by means of standardization of both variables.

As is further specified in the third column of Table 8.3, null hypothesis significance testing as implemented ini SAS PROC PLM can be used to compare a null hypothesis with one informative hypothesis specified using inequality constraints by means of a p-value in the context of generalized linear models (McCullagh and Nelder, 1989) and generalized mixed models (McCullogh and Searle, 2001). The question marks in the column labeled "Chi-bar" stress that in many situations SAS PROC PLM can also be used for the evaluation of informative hypotheses specified using simple equality constraints and effect sizes. This holds for all situations in which the user is able to provide transformations of the data such that the hypotheses of interest can are still of the form (8.10) and (8.11).

Suppose, to give a simple example, that $y_i \sim \mathcal{N}(\mu, \sigma^2)$ for $i = 1, \dots, N$ and the hypothesis of interest is $H_m : \mu > 2$. This hypothesis cannot straightforwardly be evaluated with SAS PROC PLM because it only allows hypotheses of the form $H_m : \mu > 0$, that is, the reference value must be zero. For the example at hand, this limitation can be bypassed using the transformation $y_i^* = y_i - 2$ in combination with $H_m^* : \mu^* > 0$, where μ^* denotes the mean of y_i^* for $i = 1, \dots, N$.

Informative hypotheses may be formulated using inequality and equality constraints, if the data can be transformed such that the accordingly transformed hypotheses contain only inequality constraints. Consider, again, an ANOVA with three groups and $H_m : \mu_1 > \mu_2 = \mu_3$. If the data for groups 2 and 3 are combined into group $2'$, H_m can be reformulated as $H_{m'} : \mu_1 > \mu_{2'}$. Consider also, for example, a univariate regression with three predictors $y_i = \beta_0 + \beta_1 x_{i1} + \beta_2 x_{i2} + \beta_3 x_{i3} + \epsilon_i$ and $H_m : \beta_1 > \beta_2 = \beta_3$. If the regression model is rewritten as $y_i = \beta_0 + \beta_1 x_{i1} + \beta_{2'}(x_{i2} + x_{i3}) + \epsilon_i$, the informative hypothesis can correspondingly be reformulated as $H_{m'} : \beta_1 > \beta_{2'}$.

8.4 The Order-Restricted Information Criterion

Like the Bayes factor, classical information criteria like Akaike's Information Criterion (AIC, Akaike, 1987; Burnham and Anderson, 2002) evaluate hypotheses based on their fit and complexity:

$$\text{AIC}_m = -2 \log f(\boldsymbol{Y} \mid \boldsymbol{X}, \hat{\boldsymbol{\theta}}_m, \hat{\boldsymbol{\phi}}_m) + 2p_m \tag{8.14}$$

for $H_1, \dots, H_m, \dots, H_M$. Note that $\boldsymbol{Y}$ denotes data (for example, the scores on the dependent variable in an ANOVA or multiple regression) and that $\boldsymbol{X}$ denotes ancillary data (for example, the scores on the predictors in a multiple regression). Note that the first term on the right-hand side of (8.14) is minus twice the log of the density of the data (see Appendix 10A for examples) evaluated with respect to maximum likelihood estimates of the parameters $\boldsymbol{\theta}$ and the nuisance parameters $\boldsymbol{\phi}$ that account for the constraints in H_m. In the univariate regression example with two predictors used in the previous section, $\boldsymbol{\theta} = [\beta_0, \beta_1, \beta_2]$ and $\boldsymbol{\phi} = \sigma^2$. The first term of (8.14) is often interpreted as the fit of a hypothesis: the smaller its value, the better the fit. The second term is twice the number of parameters in $\boldsymbol{\theta}$ and $\boldsymbol{\phi}$. For a multiple regression with two predictors, $p_m = 4$ because there are three regression coefficients and one residual variance. The second term is often interpreted as the size or complexity of a hypothesis. The smaller its value, the more parsimonious the hypothesis at hand.

The AIC can be interpreted as an estimate of the distance between the true hypothesis and the hypothesis under investigation. Therefore, of a set of hypotheses, the one with the smallest AIC value is supported most by the data. As is immediately clear from (8.14), the AIC cannot be used for the evaluation of informative hypotheses. Consider, for example, the hypotheses

$$H_m : \beta_1 > \beta_2 \tag{8.15}$$

and

$$H_a : \beta_1, \beta_2, \tag{8.16}$$

formulated in the context of a multiple regression model with two predictors. Although H_m is more specific and thus more parsimonious than H_a, $p_m = p_a = 4$, that is, for informative hypotheses complexity is not appropriately accounted for by the AIC. In the next two sections, two modifications of the AIC that can be used for the evaluation of informative hypotheses will be introduced. First the order-restricted information criterion (ORIC, Anraku, 1999) will be introduced; subsequently, the generalized ORIC (GORIC, Kuiper, Hoijtink, and Silvapulle (In Press); Kuiper, Hoijtink, and Silvapulle (Unpublished); Kuiper and Hoijtink (Unpublished)) will be presented.

8.4.1   ORIC

The ORIC can be used to select the best of a set of informative hypotheses specified for ANOVA. The implementation of the ORIC in `ConfirmatoryANOVA` allows evaluation of informative hypotheses of the form

$$H_m : \boldsymbol{R}_m \boldsymbol{\mu} > 0 \text{ and } \boldsymbol{S}_m \boldsymbol{\mu} = 0, \tag{8.17}$$

where each row of the $Q \times J$ matrix $\boldsymbol{R}_m$ must have exactly two non-zero elements, one with the value 1 and one with the value -1. Using these constraints, informative hypotheses consisting of (partial) orderings of the group means can be constructed. Similarly, $\boldsymbol{S}_m$ is an $L \times J$ matrix, where each row also contains exactly two non-zero elements, one with the value 1 and one with the value -1. Using these constraints, clusters of groups with the same mean can be constructed. The main hypothesis for the example from Lucas (2003) was $H_2 : \mu_3 = \mu_5 > \{\mu_1, \mu_4\} > \mu_2$. The element $\mu_3 = \mu_5$ is obtained using $\boldsymbol{S}_m = [0\ 0\ 1\ 0\ -1]$, which combined with $\boldsymbol{\mu} = [\mu_1, \mu_2, \mu_3, \mu_4, \mu_5]$ renders $0 \times \mu_1 + 0 \times \mu_2 + 1 \times \mu_3 + 0 \times \mu_4 - 1 \times \mu_5 = 0$. The element $\mu_3 > \mu_1$ is obtained via the specification of a row in $\boldsymbol{R}_m$ of the form $[-1\ 0\ 1\ 0\ 0]$, which combined with $\boldsymbol{\mu}$ renders $-1 \times \mu_1 + 0 \times \mu_2 + 1 \times \mu_3 + 0 \times \mu_4 + 0 \times \mu_5 > 0$.

The ORIC has the same functional form as the AIC:

$$\text{ORIC}_m = -2 \log f(\boldsymbol{Y} \mid \boldsymbol{D}, \hat{\boldsymbol{\mu}}_m, \hat{\sigma}_m^2) + 2p_m. \tag{8.18}$$

Note that $\boldsymbol{D}$ contains the group membership of $i = 1, \ldots, N$ persons via their scores on $j = 1, \ldots, J$ dummy variables d_{ji}, which are 1 if a person is a member of group j and 0 otherwise. Note also that $\hat{\boldsymbol{\mu}}_m$ and $\hat{\sigma}_m^2$ are maximum likelihood estimates, that is, estimates in agreement with the constraints that are used to specify H_m. Anraku (1999) derives p_m such that it can be used to quantify the complexity of informative hypotheses:

$$p_m = 1 + \sum_{j=1}^{J} P(j \mid J, V_m, H_m)j. \tag{8.19}$$

$P(\cdot)$ is a so called level probability. A level probability is the probability that the constrained maximum likelihood estimate $\hat{\mu}_m$ has j different levels if the constrained maximum likelihood estimate is computed using a vector $z = z_1, \ldots, z_J$ sampled from a population in which $\mu_1 = \ldots = \mu_J$ and covariance matrix $V_m = diag[1/N_1, \ldots, 1/N_J]$ (see Anraku (1999) for the details). With J groups, the number of levels in $\hat{\mu}_m$ is $J - L - Q_{act}$. Note that Q_{act} is the number of active constraints in $\boldsymbol{R}_m$. An active constraint is a constraint that must be enforced. If, for example, a row of $\boldsymbol{R}_m$ specifies that $\mu_1 > \mu_2$, this constraint is active if $z_1 < z_2$ because the corresponding constrained maximum likelihood estimate is $\hat{\mu}_1 = \hat{\mu}_2 = (z_1 + z_2)/2$. A constraint is inactive if $\hat{\mu}_1 = z_1$ and $\hat{\mu}_2 = z_2$.

With equal sample sizes per group, the p_m values for the following sequence of five hypotheses

$$
\begin{aligned}
&H_1 : \mu_1, \ldots, \mu_5 \\
&H_2 : \mu_1 > \mu_2, \ldots, \mu_5 \\
&H_3 : \mu_1 > \mu_2 > \mu_3, \mu_4, \mu_5 \\
&H_4 : \mu_1 > \mu_2 > \mu_3 > \mu_4, \mu_5 \\
&H_5 : \mu_1 > \mu_2 > \mu_3 > \mu_4 > \mu_5
\end{aligned}
\tag{8.20}
$$

are

$$
\begin{aligned}
&5 \\
&4 + 1/2 = 4.5 \\
&3 + 1/2 + 1/3 = 3.83 \\
&2 + 1/2 + 1/3 + 1/4 = 3.08 \\
&1 + 1/2 + 1/3 + 1/4 + 1/5 = 2.28
\end{aligned}
\tag{8.21}
$$

respectively. As can be seen, p_m accounts for the complexity of the hypotheses under investigation: the larger the number of constraints, that is, the more parsimonious the model under investigation, the smaller the value of p_m. In this sense, p_m can be seen as the expected number of parameters under constraints.

The fourth column of Table 8.3 summarizes the main features of the ORIC as implemented in ConfirmatoryANOVA: it can be used to evaluate M hypotheses specified using simple inequality and equality constraints in the context of ANOVA models.

As can be seen in the sixth column of Table 8.2, evaluation of the hypotheses from Lucas (2003) renders the smallest ORIC value for H_2; that is, of the hypotheses under investigation, H_2 is closest to the truth. Because the ORIC is not a quantification of evidence, it cannot be determined whether H_2 is much or only slightly closer to the truth than the other hypotheses under consideration. Note, however, that Burnham and Anderson (2002) introduce Akaike weights. These weights are transformations of AIC_m for $m = 1, \ldots, M$ to numbers on a scale running from 0 to 1. As for posterior model probabilities (see Section 3.3.2), these numbers are a quantification of the complexity and fit of each hypothesis under consideration. However, unlike posterior model probabilities, these numbers are not logical probabilities. Currently, such weights have not been considered in the context of the ORIC.

8.4.2 GORIC

The GORIC (Kuiper, Hoijtink, and Silvapulle, In press; Unpublished; Kuiper and Hoijtink, Unpublished) is a generalization of the ORIC in two ways: first of all, it can be used for the evaluation of informative hypotheses in the context of the multivariate normal linear model; and second, hypotheses may be of the form

$$
H_m : \boldsymbol{R}_m \boldsymbol{\theta} > \boldsymbol{r}_m \text{ and } \boldsymbol{S}_m \boldsymbol{\theta} = \boldsymbol{s}_m,
\tag{8.22}
$$

where according to the notation used in Equation (2.2) in Chapter 2 $\boldsymbol{\theta} = [\mu_{11}, \ldots, \mu_{1J}, \beta_{11}, \ldots, \beta_{1K}, \ldots, \mu_{P1}, \ldots, \mu_{PJ}, \beta_{P1}, \ldots, \beta_{PK}]$, that is, a parameter vector containing means and regression coefficients for $p = 1, \ldots, P$ dependent variables. Note that

R_m and S_m must be of full rank. Consequently, none of the constraints may be redundant, that is, implied by the other constraints, and the first part of (8.22) may not be used to specify about equality or range constraints.

The GORIC is defined by:

$$\text{GORIC}_m = -2\log f(Y \mid D, X, \hat{\mu}_m, \hat{\beta}_m, \hat{\Sigma}_m) + 2p_m. \qquad (8.23)$$

Note that Σ denotes the residual covariance matrix of the multivariate normal linear model. Note also that estimates of the parameters of the multivariate normal linear model in agreement with the constraints in H_m are denoted by $\hat{\mu}_m, \hat{\beta}_m$ and $\hat{\Sigma}_m$. As for the ORIC, p_m quantifies the size or complexity of a hypothesis and can be seen as the expected number of parameters. The interested reader is referred to Kuiper, Hoijtink, and Silvapulle (In Press; Unpublished) for the statistical foundation and elaboration of the GORIC, and to Kuiper and Hoijtink (Unpublished) for description, examples, and manual for the software package GORIC.

The fifth column of Table 8.3 summarizes the main features of the GORIC as implemented in GORIC; it can be used to evaluate M hypotheses specified using simple inequality and equality constraints, effect sizes, and constraints on combinations of parameters in the context of the multivariate normal linear model.

8.5   Discussion

This chapter introduced two approaches rooted in classical statistics for the evaluation of informative hypotheses: null hypothesis significance testing and order-restricted information criteria. As illustrated using data from Lucas (2003) and summarized in Table 8.3, from the perspective of scientists who want to evaluate informative hypotheses the differences between both methods and the Bayesian approach can be summarized as follows. Hypothesis testing can be used to evaluate one informative hypothesis. Order-restricted information criteria can additionally be used to compare nested and nonnested informative hypotheses. The Bayesian approach can do the same as order-restricted information criteria extended with the option to evaluate informative hypotheses specified using about equality and/or range constraints. Furthermore, the Bayesian approach renders the degree of evidence in the data for each of the hypotheses under investigation. An issue with the Bayesian approach (and not with both classical approaches) is the specification of (compatible) prior distributions.

Which of these methods is preferred may depend on the availability of software. An overview of the software available for each method is provided in Chapter 9. Preference may also depend on one's position in the discussion around p-values (see, for example, Raftery (1995), Wagenmakers (2007), and Van de Schoot, Hoijtink, and Romeijn (2011)), and whether one prefers the frequentist or Bayesian paradigm (Barnett, 1999; Howard, Maxwell, and Fleming, 2000; Howson and Urbach, 2006). However, this discussion will not be repeated in this book.

In addition to Bayes factors, p-values and order-restricted information criteria, there are other approaches that can be used to evaluate informative hypotheses. The Bayesian information criterion (BIC, Raftery, 1995) can be modified such that it can be used for the evaluation of informative hypotheses (Romeijn, van de Schoot, and Hoijtink, In Press). The deviance information criterion (DIC, Spiegelhalter, Best, Carlin, and van der Linde, 2002) can also be modified such that it can be used for the evaluation of informative hypotheses

(Van de Schoot, Hoijtink, Brugman, and Romeijn, Unpublished). However, both modifications are closely related to the marginal likelihood and the Bayes factor (which is the ratio of two marginal likelihoods). Therefore, because software with which both modifications can be used is not available, those interested in modifications of BIC and DIC can resort to the use of Bayes factors. Van de Schoot, Dekovic, and Hoijtink (2010) and Van de Schoot and Strohmeier (2011) use calibrated bootstrapped p-values to test a null hypothesis versus an informative hypothesis, and to test an informative hypothesis versus an unconstrained hypothesis, in the context of structural equation models (Kline, 2005). Example macros with which this approach can be implemented in Mplus (`http://www.statmodel.com/`) can be obtained from `a.g.j.vandeschoot@uu.nl`. This approach is currently being implemented in structural equation modeling package `lavaan` (`http://lavaan.ugent.be`).

8.6 Appendix 8A: Data and Command File for `ConfirmatoryANOVA`

The software package `ConfirmatoryANOVA` (see Section 9.2.1) needs a text file containing the data with the name `data.txt` as in Table 8.4 and a text file with instructions with the name `input.txt` as in Table 8.5. Both files can be prepared using a `Windows` users interface or by manually preparing both text files. The data in `data.txt` consist of two columns. The first column contains the group membership of a person, the second column the score on the dependent variable (in the example based on Lucas (2003), the dependent variable is an influence score). Note that the data in Table 8.4 are a reconstruction of the original data with the same descriptive statistics as in Table 8.1

The commands displayed in `input.txt` render, among other things, the results displayed in Table 8.2. Note that the `Windows` users interface makes the construction of `input.txt` very easy. The first line of numbers are default values that do not have to be changed for the example at hand. The second line of numbers requests that the F-bar test `Fbar=1`, the ORIC `ORIC=1`, and Bayes factors and posterior model probabilities `BMS=1` are computed. The command `Number of models to be compared=4` requests the evaluation of four hypotheses. Subsequently it is specified that the number of constraints needed for these four hypotheses are 1, 1, 6, and 5, respectively.

The constraint needed for the construction of H_0 is specified using the first line 1 2 3 4 5 from `ordering of means in restriction` and the first line 1 1 1 1 1 from `(Order) Restrictions`. Together, both lines state that $\mu_1 = \mu_2 = \mu_3 = \mu_4 = \mu_5$. The constraint needed for the construction of H_a is specified using the second line 1 2 3 4 5 from `ordering of means in restriction` and the second line 0 0 0 0 0 from `(Order) Restrictions`. Together, both lines state $H_a : \mu_1, \mu_2, \mu_3, \mu_4, \mu_5$. The last five lines from `ordering of means in restriction` and `(Order) Restrictions` specify the five constraints needed to specify $H_2 : \mu_3 = \mu_5 > \{\mu_1, \mu_4\} > \mu_2$. The line 3 5 1 2 4 combined with the line 1 1 0 0 0 states that $\mu_3 = \mu_5$. The line 5 1 2 3 4 combined with the line 1 -3 0 0 0 states that $\mu_5 > \mu_1$. From these examples it can be derived that 1 1 states that the corresponding means are equal, that 0 0 states that the corresponding means are unconstrained, and that 1 -3 states that the first mean is larger than the second mean. The last line gives default values. The interested reader is referred to Kuiper, Klugkist, and Hoijtink (2010) for elaboration. Note once more, that `input.txt` is very easy to construct using the `Windows` user interface for `ConfirmatoryANOVA`.

TABLE 8.4: ConfirmatoryANOVA Data File data.txt. Data Simulated to Have the Same Descriptives as in Table 8.1

1	3.58	2	1.67	3	1.39	4	1.57	5	1.38
1	-0.15	2	1.85	3	4.53	4	2.97	5	4.58
1	0.67	2	0.5	3	1.18	4	1.45	5	3.48
1	2.22	2	0.63	3	3.67	4	1.78	5	0.39
1	2.56	2	0.04	3	2.88	4	0.97	5	3.89
1	1.7	2	2.8	3	3.5	4	4.3	5	4.18
1	-0.45	2	0.02	3	1.74	4	5.12	5	3.72
1	1.08	2	1.32	3	5.43	4	1.67	5	4.79
1	4.83	2	-0.2	3	4.69	4	1.65	5	3.63
1	2.44	2	0.14	3	1.62	4	1.66	5	1.55
1	6.74	2	2.65	3	3.67	4	1.32	5	-0.46
1	1.4	2	0.07	3	3.7	4	4.1	5	2.04
1	-0.03	2	3	3	1.79	4	1.23	5	2.26
1	0.71	2	0.53	3	2.98	4	1.58	5	4.27
1	4.3	2	-0.38	3	6.31	4	1.94	5	5.67
1	5.47	2	0.48	3	4.9	4	4.51	5	4.69
1	2.44	2	3.87	3	2.23	4	0.89	5	2.17
1	0.81	2	1.4	3	3.71	4	0.86	5	2.27
1	3.65	2	2.02	3	3.41	4	4.81	5	4.1
1	1.41	2	2.37	3	6.84	4	0.68	5	3.74
1	-0.38	2	1.99	3	2.88	4	0.47	5	4.54
1	3.11	2	1.32	3	-0.31	4	0.56	5	1.71
1	0.21	2	0.86	3	-0.08	4	2.62	5	4.74
1	3.51	2	2.3	3	4.14	4	2.41	5	4.21
1	2.63	2	-0.28	3	3.45	4	1.49	5	4.1
1	4.74	2	0.78	3	5.2	4	5.01	5	4.8
1	2.12	2	1.48	3	2.2	4	3.94	5	2.9
1	4.36	2	3.29	3	-0.03	4	2.69	5	2.35
1	3.26	2	2.14	3	3.71	4	2.1	5	4.09
1	0.94	2	1.18	3	4.67	4	0.57	5	1.09

TABLE 8.5: `ConfirmatoryANOVA` Command File `input.txt` Rendering the Results in
Table 8.2

```
Seed       Fbar     ORIC     BMS
123        100000   100000   500000
Perform    Fbar     ORIC     BMS      (1 = yes, 0 = no)
           1        1        1
Number of models to be compared
4
Number of restrictions per model
1
1
6
5
Ordering of means in restriction
1 2 3 4 5
1 2 3 4 5
1 2 3 4 5
1 4 2 3 5
1 5 2 3 4
3 2 1 4 5
3 4 1 2 5
3 5 1 2 4
3 5 1 2 4
5 1 2 3 4
5 4 1 2 3
1 2 3 4 5
4 2 1 3 5
(Order) Restrictions
1 1 1 1 1
0 0 0 0 0
1 -3 0 0 0
1 -3 0 0 0
1 -3 0 0 0
1 -3 0 0 0
1 -3 0 0 0
1 -3 0 0 0
1 1 0 0 0
1 -3 0 0 0
1 -3 0 0 0
1 -3 0 0 0
1 -3 0 0 0
delta      pv
0          2
```

Chapter 9

Software

9.1 Introduction

In this chapter, seven software packages that can be used for the evaluation of informative hypotheses are presented. The main features of each package are presented in Table 9.1. The following abbreviations are used for software packages: CONFA denotes `ConfirmatoryANOVA`, GMD denotes `GenMVLData`, CONTIN denotes `ContingencyTable`, and WB denotes `WinBUGS`. The following abbreviations are used for statistical models: UNLM/MNLM denotes univariate/multivariate normal linear model, CONTIN denotes contingency table analysis, GLM denotes generalized linear models, GMM denotes generalized mixed models, and GEN denotes statistical models in general. The entry Y denotes that an option is available in the software package at hand. The entry ? denotes that the option will further be discussed in the section dealing with the software package at hand.

In each section the main features of each software package will shortly be discussed. Each section is concluded with a discussion of features that are essential for proper use of the software package discussed. All the software packages can be found at or via the website related to this book. New developments and software updates in the context of the evaluation of informative hypotheses will be monitored and announced at both the website related to this book, and the website on informative hypotheses maintained by the Utrecht group working on informative hypotheses. Both websites can be accessed via the home page of the author at `http://tinyurl.com/hoijtink`.

TABLE 9.1: Options Available in the Software Packages

Method	CONFA	BIEMS	GORIC	GMD	CONTIN	SAS	WB
Constraints Available for the Construction of H_m							
Simple > <	Y	Y	Y		Y	Y	Y
Simple =	Y	Y	Y		Y	?	
About Equality		Y					
Range		Y					
Effect Sizes		Y	Y			?	Y
Combinations		Y	Y		Y	Y	Y
Hypotheses to which H_m can be Compared							
H_0	Y	Y	Y		Y	Y	
H_a	Y	Y	Y		Y		Y
$H_{m'}$	Y	Y	Y		Y	Y	Y
H_{m_c}	Y						Y
Most Important Result							
F-bar	Y						
Chi-bar						Y	
ORIC	Y						
GORIC			Y				
Bayes factor	Y	Y			Y		Y
CI/SD BF		Y					Y
Simulate Data				Y			
Statistical Models, Windows Users Interface, Manuals, and Examples							
Statistical Models	ANOVA	ANOVA UNLM MNLM	ANOVA UNLM MNLM	UNLM MNLM	CONTIN	UNLM MNLM GLM MLM	GEN
Users Interface	Y	Y		Y		Y	Y
Manuals	Y	Y	Y	Y	Y		
Examples	Y	Y	Y	Y	Y		Y

9.2   Software Packages

9.2.1   ConfirmatoryANOVA

ConfirmatoryANOVA was written by Rebecca Kuiper and Irene Klugkist. A Windows user interface is available. It is free if researchers refer to Klugkist, Laudy, and Hoijtink (2005), Kuiper and Hoijtink (2010), and Kuiper, Klugkist, and Hoijtink (2010) when using this package. Chapter 8 is built around an application of ConfirmatoryANOVA. The data and command file used are presented and discussed in Appendix 8A. The main features of this package are listed in Table 9.1.

The computation of Bayes factors is, by default, based on a sample of 500,000 from both the prior and posterior distribution. This is indicated by BMS=500000 in Table 8.5 in Appendix 8A. Using the estimation procedure described in Section 10.5.1, ConfirmatoryANOVA accurately evaluates the equality constraints that are used to specify informative hypotheses. However, this procedure is not used to evaluate the inequality constraints used in the specification of informative hypotheses. ConfirmatoryANOVA therefore increases the sample size BMS if more than six means are involved in the hypotheses of interest. However, these sample sizes are smaller than the recommended sample sizes in Table 7.10 in Section 7.5.3.

It is therefore recommended to manually specify the sample size **BMS** if more than six means are involved in the hypotheses of interest.

ConfirmatoryANOVA does not give BF_{mm_c}, f_m or c_m. However, often it is not too complicated to compute c_m manually. Subsequently, $f_m = BF_{ma} \times c_m$ and $BF_{mm_c} = (f_m/c_m)/((1 - f_m)/(1 - c_m))$.

9.2.2 BIEMS

BIEMS was written by Joris Mulder. A **Windows** user interface is available. It is free if researchers refer to Mulder, Klugkist, van de Schoot, Meeus, Selfhout, and Hoijtink (2009); Mulder, Hoijtink, and Klugkist (2010); and Mulder, Hoijtink, and de Leeuw (In Press) when using this software package. Examples of the application of **BIEMS** are given in Chapters 3, 4, 5, and 6. These chapters and their appendices also contain examples of the data and command files needed to run **BIEMS**. The main features of this package are listed in Table 9.1.

It is easy to run **BIEMS** using the **Windows** user interface or by manually specifying the required command files. The next three subsections highlight features that users of **BIEMS** ought to be aware of.

9.2.2.1 Compatibility of Prior Distributions

Property 4.2 in Chapter 4 stated that the support in the data for two informative hypotheses H_m and $H_{m'}$ can only be quantified using $BF_{mm'} = BF_{ma}/BF_{m'a}$ if the unconstrained prior distributions corresponding to H_m and $H_{m'}$ are compatible, that is, identical. There are two ways to determine whether compatibility holds:

- If **BIEMS** is used to evaluate a hypothesis the restriction matrix used to compute the means of the prior distribution is displayed under *****Prior Restrictions***** at the bottom of the output. If the restriction matrix is the same for H_m and $H_{m'}$, the corresponding unconstrained prior distributions are compatible.

- A more elaborate way to determine compatibility is application of the following set of rules:

  - Replace in each hypothesis about equality constraints by the corresponding equality constraints, e.g., $|\theta_1 - \theta_2| < .5$ becomes $\theta_1 = \theta_2$.
  - Replace in each hypothesis range constraints by the corresponding equality constraints, e.g., $.2 < \theta_1 < .5$ becomes $\theta_1 = .35$.
  - Replace all remaining inequality constraints by equality constraints.

  Two hypotheses are compatible if they are the same (possibly after rewriting) after application of these three rules.

Note that the **BIEMS Windows** interface issues a warning if the prior distributions of the hypotheses under consideration are incompatible.

If prior distributions are incompatible, the **BIEMS Windows** interface will automatically choose a prior distribution that is appropriate for all hypotheses under consideration. The stand alone version of **BIEMS** can manually be instructed to use compatible prior distributions. Examples of this situation were given in Section 4.4.2 in Chapter 4 and Appendix 6C. Two approaches often render compatible prior distributions for the hypotheses under investigation.

The first approach consists of the following steps:

- Replace all about equality and range constraints in H_m and $H_{m'}$ by the corresponding equality constraint.

- Replace all inequality constraints in H_m and $H_{m'}$ by equality constraints.

- Construct an imaginary hypothesis H_* by combining the constraints remaining after the execution of the first two steps. Delete redundant constraints, for example, $\theta_1 = \theta_2, \theta_2 = \theta_3, \theta_1 = \theta_3$ is over specified, the same information is contained in $\theta_1 = \theta_2, \theta_2 = \theta_3$.

- Specify `restriction_matrix.txt` such that it corresponds to H_*. For $H_* : \theta_1 = \theta_2 = \theta_3$, this renders a restriction matrix consisting of the rows `1 -1 0` followed by `0`, and `0 1 -1` followed by `0`. These rows state that $1 \times \theta_{01} - 1 \times \theta_{02} + 0 \times \theta_{03} = 0$ and that $0 \times \theta_{01} + 1 \times \theta_{02} - 1 \times \theta_{03} = 0$, where θ_{01} denotes the prior mean of θ_1. Note that the meaning of a row of numbers in the restriction matrix has been elaborated by means of a number of examples in Chapter 4. Another example is the row `0 0 1 -1`, which ends with the number `.5`. This row states that $0 \times \theta_{01} + 0 \times \theta_{02} + 1 \times \theta_{03} - 1 \times \theta_{04} = .5$.

The second approach consists of two rules:

- If hypotheses are specified using restrictions among (combinations) of parameters, give all parameters equal prior means. This is achieved using a `restriction_matrix.txt` like the one presented in Table 4.5 in Chapter 4. This restriction matrix corresponds to the imaginary hypothesis $H_* : \theta_1 = \ldots = \theta_D$, where D denotes the number of parameters involved in the hypothesis.

- If hypotheses are specified using restriction among (combinations) of parameters, and if one or more of the parameters are constrained to be larger than 0, give all parameters equal prior means, and add for one of these parameters a line to the restriction matrix stating that its prior mean is equal to zero. A `restriction_matrix.txt` resulting from the application of this rule can be found in Table 6.21 in Chapter 6. This restriction matrix corresponds to the imaginary hypothesis $H_* : \theta_1 = \ldots = \theta_D = 0$, where D denotes the number of parameters involved in the hypothesis.

The following rules can be used to determine whether or not the `restriction_matrix.txt` corresponding to H_* solves the problem of incompatible prior distributions:

- Replace in H_m and $H_{m'}$ about equality and range constraints by the corresponding equality constraints. Subsequently, replace the remaining inequality constraints by equality constraints. Call the resulting hypotheses H_{m*} and $H_{m'*}$.

- If the constraints in H_{m*} and $H_{m'*}$ are implied by the constraints in H_*, the problem of incompatible prior distributions has successfully been solved. Note that "implied" means that H_* can be obtained via the addition of zero or more equality constraints to H_{m*} and $H_{m'*}$.

If neither of these approaches give the desired result, either the hypotheses under consideration are inherently incompatible or another approach to find H_* is needed. Examples of inherently incompatible prior distributions can be found in Section 4.5.2 and Section 5.8.1. Two hypotheses are inherently incompatible if an H_* implying the constraints in H_{m*} and $H_{m'*}$ cannot be found.

9.2.2.2 Standard Deviation of the Bayes Factor

The software package BIEMS renders the standard deviation SD of BF_{ma} in the output under the label sampling error BF. A credibility interval reflecting the Monte Carlo error in the estimation of the Bayes factor can be constructed using $BF \pm 2 \times SD$. If this interval is too large, the estimate of the Bayes factor is imprecise. The precision can be improved changing the default sample size sample size=-1 of 20,000 implemented in input_BIEMS.txt to a larger number. An illustration of this feature was given in Appendix 6D.

9.2.2.3 Applications Outside the Context of This Book

Users interested in applications of BIEMS outside the context of the examples presented in Chapters 3, 4, 5, and 6, are well-advised to evaluate the performance of BIEMS in their context using data sets generated by means of GenMVLData. There is currently little experience with the evaluation of informative hypotheses in the context of repeated measurements with time varying covariates (see Section 2.7.1). There is also little experience with restriction matrices containing rows in which the numbers do not add up to 0. The latter may, for example, occur if the informative hypothesis of interest is $H_m : 2 \times \theta_1 - \theta_2 > 0$, which renders a restriction matrix containing the row 2 -1 0.

9.2.3 GORIC

GORIC was written by Rebecca Kuiper. It is free if researchers refer to Kuiper, Hoijtink, and Silvapulle (In Press); Kuiper, Hoijtink, and Silvapulle (Unpublished), and Kuiper and Hoijtink (Unpublished) when using this package. The main features of the package are listed in Table 9.1 and are discussed in Chapter 8.

GORIC is a novel package that has not yet extensively been evaluated. The interested reader is referred to Kuiper, Nederhoff, and Klugkist (Unpublished) for evaluations of the GORIC in an ANOVA context. Researchers planning to use GORIC are well advised to use GenMVLData in combination with GORIC to evaluate the performance of the program before applying it to the informative hypotheses and data of interest. See Chapters 5 and 6 for examples of such evaluations with respect to BIEMS.

9.2.4 GenMVLData

GenMVLData was written by Joris Mulder. It can be executed as a stand alone package and from the BIEMS Windows user interface. It is free if researchers refer to De Leeuw and Mulder (Unpublished) when using this package. Examples of the application of GenMVLData are given in Chapters 5 and 6. These chapters and their appendices also contain examples of the command file needed to run GenMVLData. This package can be used to generate or simulate data from a population that is defined by the univariate or multivariate normal linear model.

9.2.5 ContingencyTable

ContingencyTable is written by Olav Laudy. It is free if researchers refer to Laudy and Hoijtink (2007), Laudy (2008), and Klugkist, Laudy, and Hoijtink (2010) when using this package. An example of the application of ContingencyTable and the data and command file needed is given in Section 7.2. The main features of the package are listed in Table 9.1.

As illustrated in Section 7.2.2, multiple runs of ContingencyTable are needed to ensure that a stable estimate of the Bayes factors of interest is obtained. In particular Bayes factors computed for informative hypotheses specified using equality constraints may be instable. Currently the best way to proceed is to use the median Bayes factor resulting from multiple runs.

`ContingencyTable` does not give BF_{mm_c}, f_m or c_m. However, often it is not too complicated to compute c_m manually. Subsequently, $f_m = BF_{ma} \times c_m$ and $BF_{mm_c} = (f_m/c_m)/((1 - f_m)/(1 - c_m))$.

9.2.6 SAS PROC PLM

`SAS PROC PLM` is part of the commercial package `SAS`. The methods implemented in SAS PROC PLM are described in Chapter 3.4 of Silvapulle and Sen (2004), at `http://www.sas.com/software/sas9/`, and in Section 8.3.2. The main features of the package are listed in Table 9.1. The two question marks that are listed were elaborated in Section 8.3.2.

9.2.7 WinBUGS

The free package `WinBUGS` and the possibilities for Bayesian inference it offers are described at `http://www.mrc-bsu.cam.ac.uk/bugs/` and in Ntzoufras (2009). The options available with respect to the evaluation of informative hypotheses are exemplified in Sections 7.3 and 7.4 and the Appendices to Chapter 7, and in Van Rossum, van de Schoot and Hoijtink (In Press). The main features of the package are listed in Table 9.1. Note that the successor of `WinBUGS` is called `OpenBugs`. The `WinBUGS` command files presented in this book can straightforwardly be run with `OpenBugs`. Interested readers are referred to `http://www.openbugs.info`.

9.3 New Developments

In the years to come, new and updated software for the evaluation of informative hypotheses will become available. These developments will be monitored at the website corresponding to this book and the website maintained by the Utrecht group working on informative hypotheses. Both websites can be accessed via the home page of the author at `http://tinyurl.com/hoijtink`.

There are currently three software packages that can potentially be used for Bayesian evaluation of informative hypotheses. Each of these packages renders a sample from the posterior distribution of the statistical model implemented. Analogous to the approach based on `WinBUGS`, this sample can be used to compute the fit f_m of an informative hypothesis. If these packages can also be used to obtain a sample from the prior distribution such that the complexity c_m of an informative hypothesis becomes available, Bayes factors can easily be computed. These packages are `Mplus`, which can be used for structural equation modeling and can be found at `http://www.statmodel.com/`; `AMOS`, which can be used for structural equation modeling and can be found at `http://www.spss.com/amos/`; and `MLwiN`, which can be used for multilevel modeling and can be found at `http://www.bristol.ac.uk/cmm/software/mlwin/`.

Van de Schoot, Dekovic, and Hoijtink (2010) and Van de Schoot and Strohmeier (2011) showed how `Mplus` can be used to compare an informative hypothesis to either the corresponding null hypothesis or the unconstrained alternative hypothesis by means of p-values computed and evaluated using a calibrated parametric bootstrap procedure. This approach opens up the possibility of evaluating informative hypotheses using p-values in the context of structural equation models. The approach is currently being implemented in the structural equation modeling package `lavaan` (http://lavaan.ugent.be).

Part IV

Statistical Foundations

This part consists of Chapter 10. In this chapter the statistical foundations of Bayesian evaluation of informative hypotheses are presented. Chapter 10 is aimed at readers who are statistically rather versed, and have an above-average interest in the statistical foundations of the Bayesian approach presented in this book.

Chapter 10

Foundations of Bayesian Evaluation of Informative Hypotheses

10.1   Introduction

In this chapter the foundations of Bayesian evaluation of informative hypotheses are presented for the multivariate normal linear model discussed in Chapters 1 through 6, and for statistical models in general as discussed in Chapter 7.

The multivariate normal linear model is given by:

$$y_{1i} = \mu_{11}d_{1i} + ... + \mu_{1J}d_{Ji} + \beta_{11}x_{1i} + ... + \beta_{1K}x_{Ki} + \epsilon_{1i}$$

$$\cdots \tag{10.1}$$

$$y_{Pi} = \mu_{P1}d_{1i} + ... + \mu_{PJ}d_{Ji} + \beta_{P1}x_{1i} + ... + \beta_{PK}x_{Ki} + \epsilon_{Pi},$$

where

$$\begin{pmatrix} \epsilon_{1i} \\ \cdots \\ \epsilon_{Pi} \end{pmatrix} \sim \mathcal{N}\left(\begin{bmatrix} 0 \\ \cdots \\ 0 \end{bmatrix}, \begin{bmatrix} \sigma_1^2 & \cdots & \sigma_{1P} \\ \cdots & \cdots & \cdots \\ \sigma_{1P} & \cdots & \sigma_P^2 \end{bmatrix} \right) = \mathcal{N}(\mathbf{0}, \boldsymbol{\Sigma}). \tag{10.2}$$

Let $\boldsymbol{\theta} = [\mu_{11}, \ldots, \mu_{1J}, \beta_{11}, \ldots, \beta_{1K}, \ldots, \mu_{P1}, \ldots, \mu_{PJ}, \beta_{P1}, \ldots, \beta_{PK}]$. For the multivariate normal linear model, the hypotheses that will be considered are of the form $H_m : \boldsymbol{R}_m\boldsymbol{\theta} > \boldsymbol{r}_m$ and $\boldsymbol{S}_m\boldsymbol{\theta} = \boldsymbol{s}_m$.

The density of the data for statistical models in general will be denoted by

$$f(\boldsymbol{Y} \mid \boldsymbol{X}, \boldsymbol{\theta}, \boldsymbol{\phi}), \tag{10.3}$$

where $\boldsymbol{Y}$ denotes the data, $\boldsymbol{X}$ denotes ancillary data that are treated as fixed, $\boldsymbol{\theta} = [\theta_1, \ldots, \theta_D]$ denotes the parameters that are subjected to inequality and/or equality constraints, and $\boldsymbol{\phi}$ denotes nuisance parameters. For the multivariate normal linear model

- $\boldsymbol{Y} = [\boldsymbol{y}_1, \ldots, \boldsymbol{y}_P]$ where $\boldsymbol{y}_p = [y_{p1}, \ldots, y_{pN}]$.

- $\boldsymbol{X} = [\boldsymbol{d}_1, \ldots, \boldsymbol{d}_J, \boldsymbol{x}_1, \ldots, \boldsymbol{x}_K]$ where $\boldsymbol{d}_j = [d_{j1}, \ldots, d_{jN}]$ and $\boldsymbol{x}_k = [x_{k1}, \ldots, x_{kN}]$.

- $\boldsymbol{\theta} = [\mu_{11}, \ldots, \mu_{1J}, \beta_{11}, \ldots, \beta_{1K}, \ldots, \mu_{P1}, \ldots, \mu_{PJ}, \beta_{P1}, \ldots, \beta_{PK}]$.

- $\boldsymbol{\phi} = \boldsymbol{\Sigma}$, that is, the covariance matrix in (10.2).

The informative hypotheses that will be considered for statistical models in general are of the form

$$H_m : \boldsymbol{R}_m t(\boldsymbol{\theta}) > \boldsymbol{r}_m \text{ and } \boldsymbol{S}_m t(\boldsymbol{\theta}) = \boldsymbol{s}_m, \tag{10.4}$$

where $t(\cdot)$ denotes a monotone transformation applied to each element of $\boldsymbol{\theta}$, or to groups of elements of the same size. Note that for the multivariate normal linear model as treated in this book, $t(\boldsymbol{\theta}) = \boldsymbol{\theta}$.

10.2   Bayes Factor

Using the notation of Chib (1995), the Bayes factor of a constrained hypothesis H_m versus the unconstrained hypothesis H_a can be written as

$$BF_{ma} = \frac{f(\boldsymbol{Y} \mid \boldsymbol{X}, \boldsymbol{\theta}, \boldsymbol{\phi})h(\boldsymbol{\theta}, \boldsymbol{\phi}|H_m)/g(\boldsymbol{\theta}, \boldsymbol{\phi}|\boldsymbol{Y}, \boldsymbol{X}, H_m)}{f(\boldsymbol{Y} \mid \boldsymbol{X}, \boldsymbol{\theta}, \boldsymbol{\phi})h(\boldsymbol{\theta}, \boldsymbol{\phi}|H_a)/g(\boldsymbol{\theta}, \boldsymbol{\phi}|\boldsymbol{Y}, \boldsymbol{X}, H_a)}. \tag{10.5}$$

The prior distribution is of the form

$$h(\boldsymbol{\theta}, \boldsymbol{\phi}|H_m) = h(\boldsymbol{\theta} \mid H_m)h(\boldsymbol{\phi}), \tag{10.6}$$

that is, $\boldsymbol{\theta}$ and $\boldsymbol{\phi}$ are independent, and only the prior distribution of $\boldsymbol{\theta}$ depends on H_m. The prior distribution of H_m can be derived from the corresponding prior distribution of H_a:

$$h(\boldsymbol{\theta}|H_m) = \frac{h(\boldsymbol{\theta}|H_a)I_{\boldsymbol{\theta} \in H_m}}{\int_{\boldsymbol{\theta}} h(\boldsymbol{\theta}|H_a)I_{\boldsymbol{\theta} \in H_m}\partial\boldsymbol{\theta}} = \frac{1}{c_m}h(\boldsymbol{\theta}|H_a)I_{\boldsymbol{\theta} \in H_m}, \tag{10.7}$$

where c_m denotes the proportion of the unconstrained prior distribution in agreement with H_m and the indicator function I equals 1 if the argument is true and 0 otherwise. Note that, c_m will be called the "complexity" of H_m. Analogously, the posterior distribution of H_m can be derived from the corresponding posterior distribution of H_a:

$$g(\boldsymbol{\theta}, \boldsymbol{\phi}|\boldsymbol{Y}, \boldsymbol{X}, H_m) = \frac{1}{f_m}g(\boldsymbol{\theta}, \boldsymbol{\phi}|\boldsymbol{Y}, \boldsymbol{X}, H_a)I_{\boldsymbol{\theta} \in H_m}, \tag{10.8}$$

where f_m denotes the proportion of the unconstrained posterior distribution in agreement with H_m. Note that f_m will be called the "fit" of H_m. If, (10.7) and (10.8) are inserted in (10.5), it reduces for every value of $\boldsymbol{\theta}$ in agreement with H_m to

$$BF_{ma} = \frac{f_m}{c_m}. \tag{10.9}$$

Furthermore, from (10.5) and (10.9) it follows that

$$BF_{mm'} = BF_{ma}/BF_{m'a} = \frac{f_m}{c_m}\Big/\frac{f_{m'}}{c_{m'}}. \tag{10.10}$$

Finally, noting that the proportion of the complement H_{m_c} of H_m in agreement with $h(\boldsymbol{\theta}|H_a)$ is $1 - c_m$ and analogously obtaining the posterior proportion $1 - f_m$, it follows that

$$BF_{mm_c} = BF_{ma}/BF_{m_ca} = \frac{f_m}{c_m}\Big/\frac{1 - f_m}{1 - c_m}. \tag{10.11}$$

These equations show that Bayes factors for the evaluation of informative hypotheses have a simple form. The interested reader is referred to Chib (1995), Carlin and Chib (1995), and Han and Carlin (2001), who present methods for the computation of the Bayes factor in situations where it does not have a simple form.

As will be elaborated in Section 10.3, the specification of the prior distribution for H_a depends on the informative hypothesis H_m at hand. This implies that (10.10) only holds if the unconstrained prior distributions corresponding to H_m and $H_{m'}$, respectively, is the same. This requirement of so-called compatible prior distributions was previously addressed in Section 4.4.2 and Property 4.2 of Chapter 4.

Note that (10.9), (10.10), and (10.11) cannot be computed for hypotheses constructed using equality constraints of the form $\boldsymbol{S}_m t(\boldsymbol{\theta}) = \boldsymbol{s}_m$ because c_m and f_m will be zero for these hypotheses. As will be elaborated in Section 10.5.1, this problem can be solved if $\boldsymbol{S}_m t(\boldsymbol{\theta}) = \boldsymbol{s}_m$ is replaced by $\lim_{\boldsymbol{\eta} \to 0} |\boldsymbol{S}_m t(\boldsymbol{\theta}) - \boldsymbol{s}_m| < \boldsymbol{\eta}$, where $\boldsymbol{\eta} = [\eta, \ldots, \eta]$. As was elaborated in Wetzels, Grasman, and Wagenmakers (2010); Klugkist, Laudy, and Hoijtink (2010); and Mulder, Hoijtink, and de Leeuw (In Press), there is an interesting analogy between (10.9) and the Savage–Dickey density ratio (Dickey, 1971; Verdinelli and Wasserman, 1995). Consider the following hypotheses: $H_0 : \mu = 0$ and $H_a : \mu \neq 0$, where $y_i \sim (\mu, 1)$ for $i = 1, \ldots, N$. Using (10.5) as the point of departure, it can immediately be seen that

$$BF_{0a} = \frac{g(\mu = 0 \mid \boldsymbol{y}, H_a)}{h(\mu = 0 \mid H_a)}, \tag{10.12}$$

because $g(\mu = 0 \mid \boldsymbol{y}, H_0) = 1$ and $h(\mu = 0 \mid H_m) = 1$. The resulting ratio of posterior density and prior density is closely related to the ratio of posterior proportion and prior proportion in (10.9). Using Wetzels, Grasman, and Wagenmakers (2010), this resemblance will be used in Section 10.5.1 to support the approach used to deal with equality constraints. The interested reader is also referred to Berger and Delampady (1987) and Sellke, Bayarri, and Berger (2001), who show that the evaluation of $H_0 : \theta = 0$ renders the same result as the evaluation of $H_{0'} : |\theta| < \eta$ for specific small values of η.

The interested reader is referred to Kass and Raftery (1995) and Lavine and Schervish (1999) for an elaborate discussion of the Bayes factor. Casella and Berger (1987), Berger and Mortera (1999), and Moreno (2005) discuss Bayes factors for the evaluation of one-sided hypotheses with respect to one parameter: $H_0 : \theta = 0$ versus $H_1 : \theta > 0$ and $H_2 : \theta < 0$. Klugkist, Laudy, and Hoijtink (2005) and Mulder, Hoijtink, and Klugkist (2010) discuss Bayes factors for general classes of constraints in the multivariate normal linear model. Related approaches in the context of contingency tables and multilevel models are presented by Klugkist, Laudy, and Hoijtink (2010) and Kato and Hoijtink (2006), respectively.

10.3   Prior Distribution

This section elaborates the specification of the prior distribution (Kass and Wasserman, 1996) if the goal is to evaluate a set of informative hypotheses. First of all, prior distributions for the multivariate normal linear model are discussed. Two situations will be distinguished: hypotheses formulated exclusively using linear inequality constraints and hypotheses formulated using equality and inequality constraints. Thereafter, the focus will change to prior distributions for statistical models in general. The two situations that will be distinguished here are hypotheses formulated exclusively using linear or nonlinear inequality constraints; and hypotheses formulated using, possibly nonlinear, equality and inequality constraints. A summary of the properties of prior distributions is presented in Table 10.1.

10.3.1   Multivariate Normal Linear Model: Inequality Constraints

10.3.1.1   Prior Distribution for Constrained Parameters

Hoijtink (Unpublished) defines the complexity of inequality constrained hypotheses using an argument based on equivalent sets of hypotheses (a formal definition of equivalent sets is given in Section 10.3.3.1). Consider, for example, the hypothesis $H_{mz} : \theta_1 > \theta_2 > \theta_3$. There are $z = 1, \ldots, 3!$ permutations of the θs rendering hypotheses that have an equivalent structure. A logical implication is to consider these hypotheses to be of the same complexity. This leads to the following definition:

Definition 10.1: An *Equivalent Set* consist of equivalent hypotheses $H_{m1}, \ldots, H_{mZ}$ for which $H_{m1} \bigcup \ldots \bigcup H_{mZ}$ encompasses 100% of the parameter space.

With $3! = 6$ equivalent hypotheses, 16.6% of the parameter space is in agreement with each hypothesis, with Z equivalent hypotheses the proportion of the parameter space in agreement with each is $1/Z$. This leads to the following definition of complexity:

Definition 10.2: The *Complexity of an Inequality Constrained Hypothesis* is the proportion of the parameter space in agreement with H_{mz}.

The following theorem is implied by the theorems proven in Hoijtink (Unpublished):

Theorem 10.1: If $h(\theta_d \mid H_a) = \mathcal{N}(\theta_0, \tau_0^2)$ for $d = 1, \ldots, D$ and H_m belongs to an equivalent set, the complexity $c_m = 1/Z$ independent of θ_0 and τ_0^2.

Note that Theorem 10.1 is a formalization of Property 3.2 presented in Section 3.2.2.2. The prior distribution specified in Theorem 10.1 is neutral with respect to the hypotheses contained in the equivalent set, and the complexity c_m is independent of the specification of the prior mean and variance.

Not all inequality constrained hypotheses belong to an equivalent set. Consider, for example,

$$H_{m1} : \quad \begin{aligned} \theta_1 - \theta_2 &> \theta_3 - \theta_4 \\ \theta_1 &> \theta_2 \\ \theta_3 &> \theta_4. \end{aligned} \tag{10.13}$$

Each element of H_{m1} can be permuted in two ways; however, although all $2 \times 2 \times 2$ permutations cover 100% of the parameter space, not each permutation is equally complex. For example, 0% of the parameter space is in agreement with

$$H_{m2} : \quad \begin{aligned} \theta_1 - \theta_2 &> \theta_3 - \theta_4 \\ \theta_2 &> \theta_1 \\ \theta_3 &> \theta_4 \end{aligned} \quad . \tag{10.14}$$

Consider also $H_{m1} : \theta_1 > \theta_2 + \theta_3$. Here the permutation argument is lost because this hypothesis does not have the same structure as $H_{m2} : \theta_2 + \theta_3 > \theta_1$. Nevertheless, it still holds that the union of H_{m1} and H_{m2} covers 100% of the parameter space. Finally consider $H_{m1} : \theta_1 > 0, \theta_2 > 0$. If the ordinate $(0,0)$ is treated as the natural midpoint of the parameter space; the permutation argument renders four hypotheses (another is $H_{m2} : \theta_1 < 0, \theta_2 < 0$) that are of equal complexity. Note that, Jeffreys (1961, Chapters 5 and 6) and Berger and Mortera (1999) in the univariate counterpart of H_{m1} use 0 as the mean of the prior distribution and thus as the natural midpoint of the parameter space.

The prior distribution used for hypotheses that belong to an equivalent set can also be used for hypothesis of the form

$$H_m : \boldsymbol{R}_m \boldsymbol{\theta} > \boldsymbol{r}_m, \tag{10.15}$$

where $\boldsymbol{R}_m$ is of full rank. Note that the full rank requirement implies exclusion of about equality constraints like $|\theta_1 - \theta_2| < \eta$, which could be constructed using

$$\boldsymbol{R}_m = \begin{bmatrix} 1 & -1 \\ -1 & 1 \end{bmatrix} \tag{10.16}$$

and

$$\boldsymbol{r}_m = [-\eta, -\eta]. \tag{10.17}$$

The following theorem (see Hoijtink (Unpublished) for a related theorem) will be proven:

Theorem 10.2: If $h(\theta_d \mid H_a) = \mathcal{N}(\theta_0, \tau_0^2)$ for $d = 1, \ldots, D$ and $\boldsymbol{R}_m$ is of full rank, c_m is independent of θ_0 for $\tau_0^2 \to \infty$.

Proof of Theorem 10.2 for Hypotheses of the Form (10.15): Let the rows of R_m be denoted by $R_{m1}, \ldots, R_{mQ}$ and $r_m = [r_{m1}, \ldots, r_{mQ}]$. Then

$$c_m = \qquad\qquad\qquad\qquad\qquad\qquad\qquad\qquad\qquad\qquad (10.18)$$

$$P(R_{m1}\theta - r_{m1} > 0, \ldots, R_{mQ}\theta - r_{mQ} > 0 \mid \begin{bmatrix} R_{m1}\theta - r_{m1} \\ \cdots \\ R_{mQ}\theta - r_{mQ} \end{bmatrix} \sim \mathcal{N}(M\theta_0 - r_m, C\tau_0^2)),$$

where

$$M = \begin{bmatrix} \sum_d R_{1d} \\ \cdots \\ \sum_d R_{Qd} \end{bmatrix} \qquad\qquad (10.19)$$

and

$$C = \begin{bmatrix} \sum_d R_{1d}^2 & \cdots & \sum_d R_{Qd}R_{1d} \\ \cdots & \cdots & \cdots \\ \sum_d R_{1d}R_{Qd} & \cdots & \sum_d R_{Qd}^2 \end{bmatrix}. \qquad\qquad (10.20)$$

Because $\lim_{\tau_0^2 \to \infty} P(A > 0 \mid A \sim \mathcal{N}(M\theta_0 - r_m, C\tau_0^2)) = P(A > 0 \mid A \sim \mathcal{N}(0, C\tau_0^2)) = P(A > 0 \mid A \sim \mathcal{N}(0, C))$, where A is a vector of length Q containing $[R_{m1}\theta - r_{m1}, \ldots, R_{mQ}\theta - r_{m1}]$, the $\lim_{\tau_0^2 \to \infty} c_m$ reduces to

$$c_m = P(R_{m1}\theta - r_{m1} > 0, \ldots, R_{mQ}\theta - r_{mQ} > 0 \mid [\begin{matrix} R_{m1}\theta - r_{m1} \\ \cdots \\ R_{mQ}\theta - r_{mQ} \end{matrix}] \sim \mathcal{N}(0, C)), \quad (10.21)$$

that is, independent of θ_0.

An implication of Theorem 10.2 for hypotheses of the form $H_m : R_m\theta > r_m$ can be illustrated using the hypothesis $H_m : \theta_1 > \theta_2 + 1 > \theta_3 + 4$. Using the prior distribution from Theorem 10.2, it can be seen that $c_m = 1/6$, that is, the complexity of the hypothesis $H_m : \theta_1^* > \theta_2^* > \theta_3^*$ where $\theta_1^* = \theta_1$, $\theta_2^* = \theta_2 + 1$ and $\theta_3^* = \theta_3 + 4$. Similarly, the complexity of $H_m : \theta > 2$ is $1/2$, that is, the complexity of the hypothesis $H_m : \theta^* > 0$, where $\theta^* = \theta - 2$. Stated otherwise, application of Theorem 10.2 to hypotheses of the form (10.15) implies that the parameter space is centered around r_m instead of 0.

10.3.1.2　Prior Distribution for the Nuisance Parameters

The prior distribution for the nuisance parameters is

$$h(\Sigma) = \text{Inv-}\mathcal{W}(\nu_0, \Sigma_0), \qquad\qquad (10.22)$$

that is, an inverse Wishart distribution . As formulated in Property 3.1 and elaborated in Section 3.7.2.1, the complexity c_m of an informative hypothesis is independent of the prior distribution for the nuisance parameters. Furthermore, if it is specified to be vague, the fit f_m is virtually independent of the prior distribution for the nuisance parameters. A vague inverse Wishart distribution is obtained using small degrees of freedom $\nu_0 = P + 2$ and, for example, $\Sigma_0 = \hat{\Sigma}$, where $\hat{\Sigma}$ denotes an estimate of the residual covariance matrix obtained under H_a. Note that, for this specification, $E(\Sigma) = \Sigma_0/(\nu_0 - P - 1) = \Sigma_0$.

10.3.1.3　Properties

The properties of the prior distribution specified in this section are summarized in the first line of the top panel in Table 10.1 (the entry "Y" denotes yes and the entry "N" denotes no).

TABLE 10.1: Prior Distributions

H_m	Priors $h(\boldsymbol{\theta}\|Ha)$	$h(\phi)$	Complexity D/O/A	Fit D/O
Priors and Properties for the Multivariate Normal Linear Model				
$\boldsymbol{R}_m\boldsymbol{\theta} > \boldsymbol{r}_m$ Option 1	$\mathcal{N}(\theta_0 = 0, \tau_0^2 \to \infty)$	Vague	Y/Y/Y	Y/Y
$\boldsymbol{R}_m\boldsymbol{\theta} > \boldsymbol{r}_m$ Option 2	$\mathcal{N}(\boldsymbol{\theta}_0^r, \boldsymbol{T}_0^r)$	Jeffreys	Y/Y/Y	Y/Y
$\boldsymbol{R}_m\boldsymbol{\theta} > \boldsymbol{r}_m$ and $\boldsymbol{S}_m\boldsymbol{\theta} = \boldsymbol{s}_m$	$\mathcal{N}(\boldsymbol{\theta}_0^r, \boldsymbol{T}_0^r)$	Jeffreys	Y/N/Y	Y/Y
Priors and Properties for Statistical Models in General				
$\boldsymbol{R}_m t(\boldsymbol{\theta}) > \boldsymbol{r}_m$				
Equivalent Sets	Symmetric	Vague	Y/Y/Y	Y/Y
Normal Prior Option 1	$\mathcal{N}(\theta_0 = 0, \tau_0^2 \to \infty)$	Vague	Y/Y/Y	Y/Y
Normal Prior Option 2	$\mathcal{N}(\boldsymbol{\theta}_0^r, \boldsymbol{T}_0^r)$	Vague	Y/Y/Y	Y/Y
Non Normal Prior	Symmetric	Vague	Y/N/Y	Y/Y
$\boldsymbol{R}_m t(\boldsymbol{\theta}) > \boldsymbol{r}_m$ and $\boldsymbol{S}_m t(\boldsymbol{\theta}) = \boldsymbol{s}_m$				
$\boldsymbol{\theta}$ Bounded	Uninformative	Vague	Y/N/Y	Y/Y
$\boldsymbol{\theta}$ Unbounded	To be developed	Vague	—	—

The prior distribution is default in the sense that only the informative hypotheses of interest must be specified. The prior distribution is objective in the sense that c_m is independent of θ_0 for $\tau_0^2 \to \infty$. The prior distribution is appropriate in the sense that $h(\boldsymbol{\theta} \mid H_a)$ is neutral with respect to hypotheses belonging to an equivalent set. This is denoted by the entries "Y" in the columns labeled D (default), O (objective), and A (appropriate) under the label Complexity. Note that the prior specification given in this section should be used if WinBUGS (see Section 9.2.7) is used to evaluate informative hypotheses specified exclusively using inequality constraints in the context of the univariate and multivariate normal linear models.

Royall (1997) has written a book with the title *Statistical Evidence. A Likelihood Paradigm*. This book is a "must-read" for every statistician. It discusses the quantification of evidence in the data for two competing hypotheses by means of the likelihood ratio. An important example is

$$LR_{mm'} = \frac{f(\boldsymbol{Y} \mid \boldsymbol{X}, \hat{\boldsymbol{\theta}}_m, \hat{\boldsymbol{\phi}}_m)}{f(\boldsymbol{Y} \mid \boldsymbol{X}, \hat{\boldsymbol{\theta}}_{m'}, \hat{\boldsymbol{\phi}}_{m'})}, \tag{10.23}$$

where $\hat{\boldsymbol{\theta}}_m$ and $\hat{\boldsymbol{\phi}}_m$ denote the maximum likelihood estimates of the parameters under H_m. The likelihood ratio cannot be used for the evaluation of inequality constrained hypotheses because it does not adequately account for model complexity. Consider, for example, $H_m : \mu_1 > \mu_2$ and $H_{m'} : \mu_1, \mu_2$, that is, hypotheses in the context of a simple two group ANOVA model. If $\bar{y}_1 > \bar{y}_2$, then $LR_{m,m'} = 1$ because parameter estimates are identical under H_m and $H_{m'}$, that is, since complexity is not accounted for, the support in the data for H_m and $H_{m'}$ is the same.

Royall (1997) devotes one chapter to Bayesian approaches. He is not enthusiastic because of the subjective element entering the analysis in the form of the prior distribution. He might however, support the Bayes factor resulting from the prior distribution described in the previous two sections because it is default, objective, based on an appropriate quantification of complexity, and the ratio of likelihoods averaged over the prior distribution of the constrained and the nuisance parameters:

$$BF_{mm'} = \frac{\int_{\boldsymbol{\theta},\phi} f(\boldsymbol{Y} \mid \boldsymbol{X}, \boldsymbol{\theta}, \phi)h(\boldsymbol{\theta} \mid H_m)h(\phi)\partial\boldsymbol{\theta}, \phi}{\int_{\boldsymbol{\theta},\phi} f(\boldsymbol{Y} \mid \boldsymbol{X}, \boldsymbol{\theta}, \phi)h(\boldsymbol{\theta} \mid H_{m'})h(\phi)\partial\boldsymbol{\theta}, \phi}. \tag{10.24}$$

10.3.2   Multivariate Normal Linear Model: All Constraints

10.3.2.1   Conjugate Expected Constrained Posterior Prior

Mulder, Hoijtink, and Klugkist (2010) do not limit themselves to hypotheses of the form $H_m : \boldsymbol{R}_m\boldsymbol{\theta} > \boldsymbol{r}_m$. They want to evaluate hypotheses of the form $H_m : \boldsymbol{R}_m\boldsymbol{\theta} > \boldsymbol{r}_m$ and $\boldsymbol{S}_m\boldsymbol{\theta} = \boldsymbol{s}_m$ and do not require that $\boldsymbol{R}_m$ is of full rank, that is, equality constraints and about equality constraints like $|\theta_1 - \theta_2| < r$ are not excluded.

The prior distribution used is called the conjugate expected constrained posterior prior (*CECPP*, the name will be explained below), further elaborated by Mulder, Klugkist, van de Schoot, Meeus, Selfhout, and Hoijtink (2009) and finalized in Mulder, Hoijtink, and de Leeuw (In Press). It has the following form:

$$h(\boldsymbol{\theta}, \boldsymbol{\Sigma} \mid H_a) = h(\boldsymbol{\theta} \mid H_a)h(\boldsymbol{\Sigma}) = \mathcal{N}(\boldsymbol{\theta}_0^r, \boldsymbol{T}_0^r)|\boldsymbol{\Sigma}|^{-(P+1)/2}, \qquad (10.25)$$

that is, Jeffreys prior is used to specify $h(\boldsymbol{\Sigma})$ such that it is vague. The superscript r stresses that the elements of $\boldsymbol{\theta}_0^r$ must be in agreement with constraints derived from the informative hypotheses under consideration, that is,

$$A : \boldsymbol{R}^*\boldsymbol{\theta}_0^r = \boldsymbol{r}^* \text{ and } \boldsymbol{S}^*\boldsymbol{\theta}_0^r = \boldsymbol{s}^*. \qquad (10.26)$$

If only one informative hypothesis is evaluated, $\boldsymbol{R}^* = \boldsymbol{R}_m$ and $\boldsymbol{S}^* = \boldsymbol{S}_m$. However, if more than one informative hypothesis has to be evaluated $\boldsymbol{R}^*$ and $\boldsymbol{S}^*$ have to contain the constraints of each hypothesis under consideration. In this way the problem of incompatible prior distributions is avoided. This issue was previously discussed in in Section 4.4.2 and Section 9.2.2.1. Note that in these sections the union of $\boldsymbol{R}^*$ and $\boldsymbol{S}^*$ is called the restriction matrix or `restriction_matrix.txt`. Consider, for example, the comparison of

$$H_m : \theta_1 > \theta_2, \theta_3, \qquad (10.27)$$

for which $\boldsymbol{R}_m = [1 \ -1 \ 0]$ with

$$H_{m'} : \theta_1, \theta_2 > \theta_3, \qquad (10.28)$$

for which $\boldsymbol{R}_{m'} = [0 \ 1 \ -1]$. The unconstrained priors corresponding to H_m and $H_{m'}$ are different. However, without violating the requirements in A, we can also use

$$\boldsymbol{R}^* = \begin{bmatrix} 1 & -1 & 0 \\ 0 & 1 & -1 \end{bmatrix} \qquad (10.29)$$

for both hypotheses. This example shows that $\boldsymbol{R}^*$-matrices (and $\boldsymbol{S}^*$-matrices) rendering compatible prior distributions for all hypotheses under consideration can often easily be constructed. Note that there are situations in which the unconstrained prior distributions corresponding to H_m and $H_{m'}$ are inherently incompatible. Examples can be found in Section 4.5.2 and Section 5.8.1.

In $\boldsymbol{r}^*$ the elements of $\boldsymbol{r}_m$ involved in about equality constraints or range constraints are replaced by the center of the range. Consider, for example, $H_m : |\theta_1 - \theta_2| < .2$, that is,

$$\boldsymbol{R}_m = \begin{bmatrix} -1 & 1 \\ 1 & -1 \end{bmatrix} \qquad (10.30)$$

and $\boldsymbol{r}_m = [-.2 \ - .2]$. Because H_m states that $\theta_1 - \theta_2$ is in the interval $[-.2, .2]$, the corresponding $\boldsymbol{r}^* = [0 \ 0]$. If H_m would have stated that $\theta_1 - \theta_2$ is in the interval $[.3, .5]$, the corresponding $\boldsymbol{r}^* = [.4 \ .4]$. If only one informative hypothesis is evaluated $\boldsymbol{r}^* = \boldsymbol{r}_m$ and $\boldsymbol{s}^* = \boldsymbol{s}_m$. However, if more than one informative hypothesis has to be evaluated, $\boldsymbol{r}^*$ and $\boldsymbol{s}^*$ have to contain the constraints of each hypothesis under consideration.

Let $\boldsymbol{\tau}_0^r$ denote the diagonal of $\boldsymbol{T}_0^r$. The superscript r has one implication for the variances $\boldsymbol{\tau}_0^r$ on the diagonal of $\boldsymbol{T}_0^r$:

$$B : \boldsymbol{R}^{**}\boldsymbol{\tau}_0^r = 0, \tag{10.31}$$

where $\boldsymbol{R}^{**}$ is constructed such that parameters that are mutually constrained in at least one of the hypotheses under consideration have the same prior variance. If, for example, $H_m : \theta_1 > \theta_2, \theta_3, \theta_4$ and $H_{m'} : \theta_1, \theta_2, \theta_3 = \theta_4$, then

$$\boldsymbol{R}^{**} = \begin{bmatrix} 1 & -1 & 0 & 0 \\ 0 & 0 & 1 & -1 \end{bmatrix}. \tag{10.32}$$

There are also constraints on the covariances in $\boldsymbol{T}_0^r$. However, because the covariances in $\boldsymbol{T}_0^r$ result from the procedure (see below and Section 10.3.2.3) used to specify the parameters of the prior distribution, these constraints are automatically satisfied and will not further be elaborated.

The specification of $\boldsymbol{\theta}_0^r$ and $\boldsymbol{T}_0^r$ is achieved using a modification of the training data based approach described in Berger and Perricchi (1996, 2004) and Perez and Berger (2002). The interested reader is also referred to O'Hagan (1995) for another training data based approach. This approach is used for the construction of prior distributions in Mulder and Hoijtink (Unpublished). The training data based procedure will further be elaborated in Section 10.3.2.3; here the three main steps are presented:

- For $w = 1, \ldots, W$, construct the training data based constrained posterior prior (CPP) such that it is in agreement with A and B:

$$CPP(\boldsymbol{\theta} \mid H_a, \boldsymbol{Y}_w, \boldsymbol{X}_w)h(\boldsymbol{\Sigma}) =$$

$$\prod_{d=1}^{D} CPP(\theta_d \mid H_a, \boldsymbol{Y}_w, \boldsymbol{X}_w)h(\boldsymbol{\Sigma}) = \tag{10.33}$$

$$\prod_{d=1}^{D} \mathcal{N}(\theta_{0dw}^r, \tau_{0dw}^r)|\boldsymbol{\Sigma}|^{-(P+1)/2}),$$

where $\boldsymbol{Y}_w, \boldsymbol{X}_w$ denotes the training data that will further be elaborated in Section 10.3.2.3.

- Average the W CPPs constructed in the previous step to obtain the expected constrained posterior prior ($ECPP$):

$$ECPP(\boldsymbol{\theta}, \boldsymbol{\Sigma} \mid H_a) =$$

$$\frac{1}{W} \sum_{w=1}^{W} CPP(\boldsymbol{\theta} \mid H_a, \boldsymbol{Y}_w, \boldsymbol{X}_w)|\boldsymbol{\Sigma}|^{-(P+1)/2}. \tag{10.34}$$

- Based upon (10.34), estimates of the mean $\boldsymbol{\theta}_0^r$ and covariance matrix $\boldsymbol{T}_0^r$ in agreement with the restrictions in A and B will be obtained. These specify a normal-Jeffreys approximation of the $ECPP$ called the conjugate expected constrained posterior prior ($CECPP$) displayed in (10.25). How these estimates are obtained is further discussed in Section 10.3.2.3.

10.3.2.2 Complexity of Inequality Constrained Hypotheses

By means of a number of examples it will now be illustrated that the complexity resulting from (10.25) is identical to the complexity resulting if the prior from Theorem 10.2 is used.

Example 10.1: If $H_m : \theta_1 > \theta_2 > \theta_3$, then

$$\boldsymbol{R}^* = \begin{bmatrix} 1 & -1 & 0 \\ 0 & 1 & -1 \end{bmatrix},$$ (10.35)

and $\boldsymbol{r}^* = [0 \ 0]$, that is, $\theta_{01w}^r = \theta_{02w}^r = \theta_{03w}^r$ (requirement A) and $\tau_{01w}^r = \tau_{02w}^r = \tau_{03w}^r$ (requirement B). If $\theta_d \sim \mathcal{N}(\theta_{0w}^r, \tau_{0w}^r)$ for $d = 1, 2, 3$ and $w = 1, \ldots, W$, then $c_m = 1/6$ for any value of θ_{0w}^r and τ_{0w}^r. The same complexity is obtained using the prior distribution from Theorem 10.2.

Example 10.2: For $H_m : \theta_1 > 2, \theta_2 > 2$, it follows that $\theta_{01w}^r = 2, \theta_{02w}^r = 2$, and that $\tau_{01w}^r \neq \tau_{02w}^r$. If $\theta_1 \sim \mathcal{N}(2, \tau_{01w}^r)$ and $\theta_2 \sim \mathcal{N}(2, \tau_{02w}^r)$ for $w = 1, \ldots, W$, then $c_m = 1/4$ for any value of τ_{01w}^r and τ_{02w}^r. The same complexity is obtained using the prior distribution from Theorem 10.2.

Example 10.3: For $H_m : \theta_1 > 2\theta_2 > 3\theta_3$, it follows that $\theta_{01w}^r = 2\theta_{02w}^r = 3\theta_{03w}^r$ (requirement A) and $\tau_{01w}^r = \tau_{02w}^r = \tau_{03w}^r$ (requirement B). If $\theta_1 \sim \mathcal{N}(\theta_{0w}^r, \tau_{0w}^r)$, $\theta_2 \sim \mathcal{N}(.5 \times \theta_{0w}^r, \tau_{0w}^r)$ and $\theta_3 \sim \mathcal{N}(.33 \times \theta_{0w}^r, \tau_{0w}^r)$, then $c_m = 1/6$ for all values of θ_{0w}^r and τ_{0w}^r. The same complexity is obtained using the prior distribution from Theorem 10.2.

If the goal is to evaluate informative hypotheses exclusively specified using inequality constraints, the prior distribution (10.25) has the following properties. In line with Property 3.1 from Chapter 3, the prior distribution for $\boldsymbol{\Sigma}$ was specified such that it is vague. In line with property 3.2 from Chapter 3, the prior distribution for $\boldsymbol{\theta}$ was specified such that it is neutral with respect to hypotheses belonging to equivalent sets, and, as was illustrated using three examples, independent of the specification of the mean and variance of the prior distribution. As is summarized in the second line of the top panel in Table 10.1, this renders a prior distribution that is default, objective, and appropriate. Note that the prior specification given in this section is used if the software package BIEMS (see Section 9.2.2) is used to evaluate informative hypotheses specified exclusively using inequality constraints in the context of the univariate and multivariate normal linear models.

10.3.2.3 Elaboration for Inequality and Equality Constrained Models

As was elaborated in the previous section, as long as the parameters $\boldsymbol{\theta}_0^r$ and $\boldsymbol{T}_0^r$ of the *CECPP* are in agreement with the constraints in A and B, the complexity c_m of hypotheses $H_m : \boldsymbol{R}_m \boldsymbol{\theta} > \boldsymbol{r}_m$ with a full rank $\boldsymbol{R}_m$ is independent of the values of $\boldsymbol{\theta}_0^r$ and $\boldsymbol{T}_0^r$. This is not the case for hypotheses of the form $H_m : \boldsymbol{R}_m \boldsymbol{\theta} > \boldsymbol{r}_m$ and $\boldsymbol{S}_m \boldsymbol{\theta} = \boldsymbol{s}_m$ without the requirement that $\boldsymbol{R}_m$ is of full rank. For hypotheses formulated using equality constraints and/or about equality constraints, it holds that BF_{ma} increases if the prior variance(s) increase (see Property 3.3 in Chapter 3). This phenomenon is known as the Lindley–Bartlett paradox (Lindley, 1957). This paradox was discussed in Section 3.7.2.2 of Appendix 3B.

To avoid the Lindley–Bartlett paradox, the parameters of the prior distribution are determined using training data (Berger and Perricchi, 1996, 2004; Perez and Berger, 2002), that is, chosen such that the information in the prior is consistent with the information in the smallest subset of the data that allows estimation of these parameters. This approach was illustrated in Section 3.6.2.2 of Appendix 3A where the parameters of the prior distribution for an ANOVA model were specified. Here the main steps of the specification of the parameters of the *CECPP* displayed in (10.25) as presented in Section 10.3.2.1 will now further be elaborated. The interested reader is referred to Mulder, Hoijtink, and Klugkist (2010); Mulder, Klugkist, van de Schoot, Meeus, Selfhout, and Hoijtink (2009); and Mulder, Hoijtink, and de Leeuw (In Press) for detailed information and elaborations.

The following procedure summarizes the steps that must be taken in order to specify the values of the parameters of the *CECPP*:

- Set the training sample size at $J + K + 1$.

- Execute the following five steps for $w = 1, \ldots, W$ (in the software package BIEMS introduced in Section 9.2.2, the number of training samples W is fixed at 1,000):

  1. Randomly sample $J + K + 1$ cases from the total of N. Denote these cases by $\boldsymbol{Y}_w, \boldsymbol{X}_w$. Note that this corresponds to $(J + K + 1) \times P$ observations, which for $P > 1$ is more than the minimum number of observations necessary in order to be able to execute Step 2 below. Note also, that it is necessary to increase the training sample size if the diagonal matrix resulting from Step 3 below has values close to zero (Mulder, Hoijtink, and de Leeuw, In Press). This may happen if, for example, the dependent variable is measured on a Likert scale with five categories. This will inevitably result in training samples, in which each case has the same score on the dependent variable.

  2. Maximize the density of the data $f(\boldsymbol{Y}_w \mid \boldsymbol{X}_w, \boldsymbol{\theta}^r_{0w}, \boldsymbol{\Sigma}_w)$ of the multivariate normal linear model in (10.1) and (10.2) with respect to $\boldsymbol{\theta}^r_{0w}$ and a diagonal matrix $\boldsymbol{\Sigma}_w$, subject to the constraints in A.

  3. Estimate the asymptotic covariance matrix of $\hat{\boldsymbol{\theta}}^r_{0w}$ assuming that these estimates are obtained subject to the constraints in B, and diagonalize the result.

  4. $\boldsymbol{T}^r_{0w}$ is obtained if the covariance matrix resulting from the previous step is scaled such that it corresponds to a training sample of $J + K - Q - L + P$ observations instead of $(J + K + 1) \times P$ observations, which corresponds to the number of regression parameters minus the number of constraints plus the size of the diagonal of $\boldsymbol{\Sigma}_w$. Note that, $J + K - Q - L + P$ observations is minimum number of observations necessary for the estimation of $\boldsymbol{\theta}^r_{0w}$ and a diagonal matrix $\boldsymbol{\Sigma}_w$.

- Sample 1,000 vectors $\boldsymbol{\theta}$ from each of the $w = 1, \ldots, W$ posterior priors.

- Use the sample of $1,000 \times W$ vectors $\boldsymbol{\theta}$ to compute, if possible, robust estimators of $\boldsymbol{\theta}^r_0$ (Huber, 1981, Chapter 3.2, the median) and $\boldsymbol{T}^r_0$ (Huber, 1981, Chapter 8.2 using the median absolute deviation for the estimation of the variances involved), that are in agreement with A and B (Mulder, Hoijtink and de Leeuw (In Press) elaborate when additional smoothing of the estimators is required). Because training samples are rather small, one or more of them may contain observations that are not representative for the sample as a whole. As result the set $\boldsymbol{\theta}_{0w}, \boldsymbol{T}_{0w}$ for $w = 1, \ldots, W$ may contain elements that are outliers in the sense that $\boldsymbol{\theta}^r_0$ and $\boldsymbol{T}^r_0$ are to a large extend determined by these outliers. Using robust estimators is an effective way to deal with this problem. See Mulder, Hoijtink and de Leeuw (In Press) for an elaboration of the situations in which robust estimators can be obtained.

The procedure described in this section is tuned to specification of the prior distribution of the parameters that are subjected to constraints. It is therefore important to note that the μs in (10.1) are only adjusted means if the xs, that is, the covariates, are standardized.

Furthermore, in a multiple or multivariate regression the constraints are only imposed on standardized regression coefficients, if both the ys and the xs are standardized.

The properties of the prior distribution specified in this section are summarized in the third line of the top panel in Table 10.1. The prior distribution is default and appropriate. However, it is not objective because the complexity of informative hypotheses specified using equality, about equality, or range constraints depends on the specification of the prior distribution. Note that the prior specification given in this section is in line with Property 3.4 from Chapter 3 and is used if the software package BIEMS (see Section 9.2.2) is used to evaluate informative hypotheses specified using equality and inequality constraints in the context of the univariate and multivariate normal linear models.

10.3.3   Statistical Models in General: Inequality Constraints

This section deals with statistical models of the form

$$f(\mathbf{Y} \mid \mathbf{X}, \boldsymbol{\theta}, \boldsymbol{\phi}), \tag{10.36}$$

and informative hypotheses of the form

$$H_m : \mathbf{R}_m t(\boldsymbol{\theta}) > \mathbf{r}_m \text{ and } \mathbf{S}_m t(\boldsymbol{\theta}) = \mathbf{s}_m, \tag{10.37}$$

where $t(\cdot)$ is a monotone function applied to each element of the vector $\boldsymbol{\theta}$ or applied to groups of elements of the same size. Three situations are discussed: equivalent sets, models with a normal prior distribution for $\boldsymbol{\theta}$, and models with a nonnormal prior distribution for $\boldsymbol{\theta}$.

10.3.3.1   Equivalent Sets

In Section 10.3.1.1, definitions with respect to and an example of a hypothesis belonging to an equivalent set were given. Now a formal definition is provided. Let $\mathbf{R}_m = [\mathbf{R}_{m1}, \dots, \mathbf{R}_{mQ}]$ and $\mathbf{R}_{mq} = R_{mq1}, \dots, R_{mqD}$. A hypothesis of the form $H_m : \mathbf{R}_m t(\boldsymbol{\theta}) > \mathbf{0}$ is a member of an equivalent set if it has the following characteristics:

- $t(\cdot)$ is a monotone transformation applied to each element of the vector $\boldsymbol{\theta}$, or to groups of elements of the same size.

- The transformation $t(\boldsymbol{\theta})$ renders a vector $\boldsymbol{\theta}^*$ with length $D^* \in \{2, \dots, D\}$.

- To obtain $\boldsymbol{\theta}^*$, the vector $\boldsymbol{\theta}$ is divided into D^* nonoverlapping subsets of the same size. Subsequently, the same transformation $t(.)$ is applied to each subset.

- Each $R_{mqd} \in \{-1, 0, 1\}$.

- $\sum_{d=1}^{D^*} R_{mqd} = 0$ for $q = 1, \dots, Q$.

- $\mathbf{R}_{m1}$ can be divided into subsets of the same size, such that $\mathbf{R}_{mq}$ is a permutation of these subsets for $q = 2, \dots, Q$.

Definition 10.3: A hypothesis H_m is a member of an equivalent set if it is in agreement with the six characteristics listed above.

In addition to $H_m : \theta_1 > \theta_2 > \theta_3$ and $H_m : \theta_1 > \{\theta_2, \theta_3\}$, the following hypotheses are also examples of hypotheses belonging to equivalent sets. Consider a contingency table with cell probabilities $\boldsymbol{\theta} = [\pi_1, \ldots, \pi_D]$ (see Section 7.2 for an elaboration of the evaluation of informative hypotheses in the context of contingency tables). If $t(\theta_d) = \log \theta_d = \theta_d^*$, then

$$H_m : \frac{\pi_1 \pi_2}{\pi_3 \pi_4} > 1, \tag{10.38}$$

a hypothesis addressing the size of an odds ratio, is equivalent to

$$H_m : \theta_1^* + \theta_2^* - \theta_3^* - \theta_4^* > 0, \tag{10.39}$$

which is in agreement with Definition 10.3 and thus a member of an equivalent set.

Using $\boldsymbol{\theta} = [\boldsymbol{\theta}_1, \boldsymbol{\theta}_2, \boldsymbol{\theta}_3]$, where $\boldsymbol{\theta}_1 = [\pi_1, \pi_2]$, $\boldsymbol{\theta}_2 = [\pi_3, \pi_4]$, and $\boldsymbol{\theta}_3 = [\pi_5, \pi_6]$, the transformation $\theta_1^* = t(\boldsymbol{\theta}_1) = \pi_1/(\pi_1 + \pi_2)$ and analogous transformations for θ_2^* and θ_3^* can be used to formulate

$$H_m : \theta_1^* > \theta_2^* > \theta_3^*, \tag{10.40}$$

which is also in agreement with Definition 10.3 and thus a member of an equivalent set.

Theorem 10.3 below is a straightforward generalization of Theorem 1 from Hoijtink (Unpublished) and will therefore be presented without proof:

Theorem 10.3: Let

$$h(\boldsymbol{\theta}, \boldsymbol{\phi} \mid H_a) = h(\boldsymbol{\theta} \mid H_a) h(\boldsymbol{\phi}). \tag{10.41}$$

If

$$h(\theta_1, \ldots, \theta_D \mid H_a) = h(\theta_{(1)}, \ldots, \theta_{(D)} \mid H_a) \tag{10.42}$$

for all permutations $(1), \ldots, (D)$ of $1, \ldots, D$, then

$$h(\theta_1^*, \ldots, \theta_{D^*}^* \mid H_a) = h(\theta_{(1)}^*, \ldots, \theta_{(D^*)}^* \mid H_a) \tag{10.43}$$

for all permutations $(1), \ldots, (D^*)$ of $1, \ldots, D^*$. Therefore, c_m is the inverse of the size Z of an equivalent set for each hypothesis in that equivalent set.

Note that priors with the property described in Theorem 10.3 will be called symmetric prior distributions. Examples of such symmetric distributions are a multivariate normal prior distribution as in Equation (4.4) in Section 4.2.2 that has the same prior mean and variance for each mean and the same prior covariance for each pair of means, and a Dirichlet distribution as in Section 7.2.4 that has the same prior parameter for each cell probability. Note that the number of elements in an equivalent set can be determined as follows:

1. Obtain all $Q!$ permutations of the Q subsets in $\boldsymbol{R}_{m1}$.

2. Denote the number of permutations for which $\boldsymbol{R}_m \boldsymbol{\theta}^*$ is in agreement with H_m by B.

3. Then $Z = Q!/B$.

The two examples presented in this section addressed the parameters $[\pi_1, \ldots, \pi_D]$ of a contingency table. Because both hypotheses are members of an equivalent set, the Dirichlet prior distribution $\mathcal{D}(\alpha_0, \ldots, \alpha_0)$, which is in agreement with the requirement formulated in Theorem 10.3, will render $c_m = 1/Z$ for both hypotheses, where Z is the size of the set of equivalent hypotheses to which H_m belongs. For the hypothesis displayed in (10.39), $Z = 2$. For the hypothesis displayed in (10.40), $Z = 6$.

10.3.3.2   Parameterizations with Normal Prior Distributions

In this section models are discussed that can be parameterized such that a normal prior distribution can be used for $\boldsymbol{\theta}$ and $H_m : \boldsymbol{R}_m\boldsymbol{\theta} > \boldsymbol{r}_m$, where $\boldsymbol{R}_m$ is a $Q \times D$ matrix of full rank. The results presented in Section 10.3.1 are directly applicable to this situation and hold for hypotheses that are and are not members of an equivalent set.

Among the models with a normal prior distribution for $\boldsymbol{\theta}$ are, first of all, the multi-variate normal linear model that has been discussed in Section 10.3.1. Furthermore, the multilevel model (Kato and Hoijtink, 2006; Section 7.3) has a normal prior distributions for the elements of $\boldsymbol{\theta}$. Models that have not yet been considered in the context of informative hypotheses are, for example, the logistic regression model (Hosmer and Lemeshow, 2000), generalized linear models (McCullagh and Nelder, 1989), and generalized mixed models (McCullogh and Searle, 2001) with a normal prior for the regression coefficients.

10.3.3.3   Parameterizations with Nonnormal Prior Distributions

This section contains a discussion of statistical models with parameters for which a normal prior distribution is inappropriate. Examples are the cell probabilities in a contingency table and the class-specific probabilities in a latent class model.

Consider hypotheses of the form $H_m : \boldsymbol{R}_m t(\boldsymbol{\theta}) > \boldsymbol{r}_m$ with the following properties:

- The transformation $t(\boldsymbol{\theta})$ is monotone, that is, $t(\cdot)$ has a monotone relation with each (group of) element(s) of $\boldsymbol{\theta}$ upon which it is based.

- The transformation $t(\boldsymbol{\theta})$ renders a vector $\boldsymbol{\theta}^*$ with length $D^* \in 1,\ldots,D$.

- $\boldsymbol{R}_m$ is of full rank.

Consider four binomial processes (see Appendix 10A) with success probability π_d for $d = 1,\ldots,4$.

Example 10.4: If

$$H_m : \pi_1 - \pi_2 > \pi_2 - \pi_3 > \pi_3 - \pi_4 \tag{10.44}$$

and $h(\pi_d) = \mathcal{B}(a_0, b_0)$ for $d = 1,\ldots,4$, that is, a Beta distribution (see Appendix 10B) with parameters a_0 and b_0, then $c_m = .139$ if $a_0 = b_0 = 1$, $c_m = .096$ if $a_0 = 4$ and $b_0 = 1$, and $c_m = .116$ if $a_0 = b_0 = .1$. As can be seen, c_m is *not* independent of the parameters of the prior distribution.

Consider a 2×3 contingency table, with cell probabilities $\pi_1, \pi_2, \pi_3, \pi_4$ for the first row and $\pi_5, \pi_6, \pi_7, \pi_8$ for the second row.

Example 10.5: If

$$H_m : \frac{\pi_1\pi_6}{\pi_2\pi_5} > \frac{\pi_2\pi_7}{\pi_3\pi_6} > \frac{\pi_3\pi_8}{\pi_4\pi_7} \tag{10.45}$$

and $h(\boldsymbol{\pi}) = \mathcal{D}(a_0,\ldots,a_0)$, that is a Dirichlet distribution with parameters a_0 (see Appendix 10B), then $c_m = .131$ for $a_0 = .1$, $c_m = .133$ for $a_0 = 1$, and $c_m = .134$ for $a_0 = 10$. Again, c_m is not independent of the parameters of the prior distribution.

Example 10.6: If

$$H_m : \frac{\pi_1}{\pi_1 + \pi_5} - \frac{\pi_2}{\pi_2 + \pi_6} > \frac{\pi_2}{\pi_2 + \pi_6} - \frac{\pi_3}{\pi_3 + \pi_7} > \frac{\pi_3}{\pi_3 + \pi_7} - \frac{\pi_4}{\pi_4 + \pi_8},$$ (10.46)

and $h(\boldsymbol{\pi}) = \mathcal{D}(a_0, \dots, a_0)$, then $c_m = .159$ for $a_0 = .1$, $c_m = .139$ for $a_0 = 1$, and $c_m = .134$ for $a_0 = 10$. A final example where c_m is not independent of the prior distribution.

What can be concluded from these examples is that, in general, for hypotheses that do not belong to an equivalent set and do not correspond to a statistical model with a normal prior distribution for each element of $\boldsymbol{\theta}$, the measure of complexity is not independent of the specification of the prior distribution, even if the (marginal) prior distribution is the same for each element of $\boldsymbol{\theta}$. This means that in these situations, complexity is not measured objectively. For the situation sketched in Example 10.4, currently the best advice is to use a uniform prior distribution for each binomial probability. A uniform distribution is obtained using a Beta distribution with parameters $a_0 = b_0 = 1$ (see binomial distribution in Appendix 10B). This would ensure that the corresponding posterior distribution is determined completely by the data. For the situations sketched in Examples 10.5 and 10.6, a posterior distribution determined completely by the data is obtained using a Dirichlet prior distribution with $a_0 = 1$ (see contingency tables in Appendix 10B).

10.3.3.4   Prior Distribution for the Nuisance Parameters

As has been illustrated in Section 3.7.2.1 in Appendix 3B, if $h(\boldsymbol{\theta}, \boldsymbol{\phi} \mid H_a) = h(\boldsymbol{\theta} \mid H_a)h(\boldsymbol{\phi})$, c_m is independent of the specification of $h(\boldsymbol{\phi})$. If additionally this distribution is vague, the data will dominate the prior, and consequently the posterior distribution will be almost completely determined by the data. Stated otherwise, Bayes factors (which are a function of c_m and f_m) are almost independent of the specification of $h(\boldsymbol{\phi})$ as long as it is vague. An example is the use of a $\mathcal{D}(1, \dots, 1)$ distribution for the class weights in a latent class model (see Section 7.4 for further elaboration).

10.3.3.5   Summary

This section summarizes the properties of Bayes factors for the evaluation of informative hypotheses in the context of statistical models in general when the prior distribution is in agreement with the requirement formulated in Theorem 10.3. As can be seen in the first four lines of the bottom panel in Table 10.1, the complexity resulting from these prior distributions is default, objective, and appropriate. An exception is the situation with nonnormal prior distributions for the parameters of a statistical model. There, the objectivity is lost because the complexity depends on the specification of the parameters of the prior distribution.

The prior distribution used for the WinBUGS (see Section 9.2.7) illustration in the context of a multilevel model (see Section 7.3) consisted of a normal distribution for each random effects mean. Each normal distribution had a mean of zero and a variance approaching infinity. This is an example of a parametrization with a normal prior distribution, as discussed in Section 10.3.3.2. Vague specifications for the prior distributions of the nuisance parameters were used.

The prior distribution for the cell probabilities implemented in the software package ContingecyTable (see Section 9.2.5) is a $\mathcal{D}(1, \dots, 1)$ (see Appendix 10C). See Section 7.2 for an illustration. This is one example of a parametrization with a nonnormal prior distribution, as discussed in Section 10.3.3.3. Note that in the context of contingency tables, there are no nuisance parameters. Another example are the $\mathcal{B}(1, 1)$ (see Appendix 10C), that is, uniform, prior distributions for the class-specific probabilities in the latent class model (see Section

7.4). In the context of latent class models, the class weights are nuisance parameters. A vague prior specification for the nuisance parameters is obtained using a $\mathcal{D}(1, \ldots, 1)$. For both examples, the complexity of a hypothesis is measured objectively only if the hypothesis at hand is a member of an equivalent set (see the discussion in Section 10.3.3.1).

10.3.4   Statistical Models in General: All Constraints

The amount of research with respect to the specification of prior distributions if the goal is to evaluate informative hypotheses of the form $H_m : \boldsymbol{R}_m t(\boldsymbol{\theta}) > \boldsymbol{r}_m$ and $\boldsymbol{S}_m t(\boldsymbol{\theta}) = \boldsymbol{s}_m$ in contexts other than the multivariate normal linear model is rather limited. The interested reader is referred to Kato and Hoijtink (2006) for an approach implemented in the context of multilevel models that is based on the approach used by Klugkist, Laudy, and Hoijtink (2005) presented in Section 3.6.2.1 in Appendix 3A. Furthermore, Klugkist, Laudy, and Hoijtink (2010) in the context of informative hypotheses with respect to the cell probabilities of contingency tables, take advantage of the fact that the domain of the parameters of interest is constrained to the interval [0,1] and specify a Dirichlet distribution such that a priori each combination of parameter values is equally likely.

Analogous to the previous section, currently the best manner to deal with this situation is to specify the prior distribution such that

- It is in agreement with Theorem 10.3, that is, appropriate for hypotheses that belong to an equivalent set.

- If the domain of each parameter subjected to constraints does *not* have a natural lower and upper bound, develop an approach analogous to one of the two approaches illustrated in Section 3.6.2 in Appendix 3A. Note that one of these approaches uses the $CECPP$, which was elaborated in Section 10.3.2.1.

- If the domain of each parameter subjected to constraints does have a natural lower and upper bound, and the parameters are mutually independent (for example, the probabilities of different binomial processes), use a uniform prior distribution over this domain. With natural bounds, but without independence (for example, the cell probabilities of a contingency table have to add up to 1.0), use a distribution that a priori assigns the same density to each combination of parameter values.

The elaboration in Section 10.3.3.4, with respect to the nuisance parameters, also applies to the situation in which informative hypotheses are formulated using equality, about equality, and inequality constraints. All these considerations are summarized in the last two lines of the bottom panel of Table 10.1. If $\boldsymbol{\theta}$ is bounded, one can use uninformative prior distributions that render default and appropriate complexity measures. However, these complexity measures are only objective if the hypothesis of interest is a member of an equivalent set. An example of such a prior is the $\mathcal{D}(1, \ldots, 1)$ distribution (see Appendix 10C) for the cell probabilities of a contingency table as is implemented in `ContingencyTable` (see Section 9.2.5). If $\boldsymbol{\theta}$ is unbounded, prior distributions must be developed for each new situation.

10.4   Posterior Distribution

The posterior distribution is proportional to the product of the density of the data and the prior distribution:

$$g(\boldsymbol{\theta}, \boldsymbol{\phi} \mid \boldsymbol{Y}, \boldsymbol{X}, H_m) \propto f(\boldsymbol{Y} \mid \boldsymbol{X}, \boldsymbol{\theta}, \boldsymbol{\phi})h(\boldsymbol{\theta} \mid H_m)h(\boldsymbol{\phi}). \qquad (10.47)$$

In Appendix 10A the density of the data of each statistical model discussed in this book is displayed. In Appendix 10B the prior distributions that are used in this book are displayed for each statistical model. In Appendix 10C an overview of the distributions used in Appendix 10B is given.

The parametric form of the prior distribution is default and the density of the data is a given for each statistical model. Consequently, the posterior distribution, which is proportional to the product of prior and density of the data, and thus f_i are default. Furthermore, as was previously stated in Property 3.6 in Chapter 3, and further elaborated in Section 3.7.2.2, the posterior distribution is mainly determined by the data and only slightly affected by the prior distribution. This implies that f_i is virtually objective, that is, mainly determined by the data. A summary of these properties for various posterior distributions can be found in the last two columns of Table 10.1. Note that Bayes factor computed for all the entries in Table 10.1 inherit the properties of the complexity and fit measures upon which they are based. Stated otherwise, Bayes factors have the same properties as the prior upon which they are based.

10.5   Estimation of the Bayes Factor

10.5.1   Decomposition of the Bayes Factor

As can be seen in Equations (10.9), (10.10), and (10.11), $BF_{ma} = f_m/c_m$ is the central quantity for the computation of various Bayes factors. A simple strategy for the computation of BF_{ma} is

- Obtain a large sample indexed $t = 1, \ldots, T$ from $g(\boldsymbol{\theta}, \boldsymbol{\phi} \mid \boldsymbol{Y}, \boldsymbol{X}, H_a)$ and estimate f_m by means of the proportion of $\boldsymbol{\theta}^t$ in agreement with H_m.

- Obtain a large sample indexed $t = 1, \ldots, T$ from $h(\boldsymbol{\theta} \mid H_a)$ and estimate c_m by means of the proportion of $\boldsymbol{\theta}^t$ in agreement with H_m.

In problems where D is relatively small and H_m is formulated using a relatively small number of inequality constraints, the simple strategy will work fine. However, in larger problems and/or when hypotheses are also formulated using equality constraints, this strategy will break down. Consider, for example, $H_m : \mu_1 > \ldots > \mu_{10}$. For this hypothesis, because there are 10! ways in which 10 parameters can be ordered, $c_m = 1/10!$, that is, a really small number. In order to be able to accurately estimate this number using the simple strategy, according to Table 7.10 in Section 7.5.3, a sample of more than 20 million parameter vectors from prior and posterior distribution is needed. This may take so much time that evaluation of H_m is not feasible. Consider also $H_m : \mu_1 = \mu_2$. As was elaborated before, the simple strategy cannot be used when this hypothesis has to be evaluated, because the proportion of prior and posterior distribution for which two means are exactly equal is zero.

Both these problems can be addressed by means of a decomposition of the Bayes factor. To start with the latter: each equality constraint can be reformulated as $H_m : |S_m\theta - s_m| < \eta$ with $\eta = [\eta, \ldots, \eta]$ and $\eta \to 0$. This can be used to decompose the Bayes factor as follows:

$$BF_{ma} = BF_{m_{\eta_0}, m_{\eta_1}} \times BF_{m_{\eta_1}, m_{\eta_2}} \times \ldots \times BF_{m_{\eta_Z}, a}, \tag{10.48}$$

where m_{η_z} for $z = Z, \ldots, 0$ denotes the hypothesis $H_{m_{\eta_z}} : |S_m\theta - s_m| < \eta_z$, $\eta_z = .5\eta_{z+1}$, and η_0 is so small that $BF_{m_{\eta_0}, m_{\eta_1}} = 1$. Wetzels, Grasman, and Wagenmakers (2010) show that the approach presented in (10.48) is identical to the corresponding Savage–Dickey ratio if the limiting process in numerator and denominator is the same. This can be exemplified using the setup presented around (10.12). Consider again the hypotheses: $H_0 : \mu = 0$ and $H_a : \mu \neq 0$, where $y_i \sim (\mu, 1)$ for $i = 1, \ldots, N$. Using l'Hopital's rule for limits that approach $0/0$, it can be seen that

$$BF_{0a} = \lim_{\eta \to 0} \frac{\int_{-\eta}^{\eta} g(0 + \xi \mid y, H_a)\partial\xi}{\int_{-\eta}^{\eta} h(0 + \xi \mid H_a)\partial\xi} \tag{10.49}$$

$$= \lim_{\eta \to 0} \frac{g(0 + \eta/2 \mid y, H_a)/2 + g(0 - \eta/2 \mid y, H_a)/2}{h(0 + \eta/2 \mid H_a)/2 + h(0 - \eta/2 \mid H_a)/2}$$

$$= \frac{g(0 \mid y, H_a)}{h(0 \mid H_a)}.$$

The interested reader is referred to Wetzels, Grasman, and Wagenmakers (2010) for more encompassing and general derivations, and to Van Wesel, Hoijtink, and Klugkist (In Press) for a simulation study providing numerical support for a generalization of (10.49) to the context of ANOVA models. The interested reader is also referred to Berger and Delampady (1987) and Sellke, Bayarri, and Berger (2001) who show that the evaluation of $H_0 : \theta = 0$ renders the same result as the evaluation of $H_{0'} : |\theta| < \eta$ for specific small values of η.

In a similar way, the Bayes factor of a hypothesis formulated using inequality constraints versus the unconstrained hypothesis can be decomposed:

$$BF_{ma} = BF_{m_Q, m_{Q-2}} \times BF_{m_{Q-2}, m_{Q-4}} \times \ldots \times BF_{m_2, a}, \tag{10.50}$$

where m_{Q-v} denotes a hypothesis constructed using the constrains in the first $Q - v$ rows of R_m.

Note that $H_m : |S_m\theta - s_m| < \eta_z$ is equivalent to $H_m : -S_m\theta + s_m > -\eta_z, S_m\theta - s_m > -\eta_z$. This implies that each equality constrained hypothesis can be represented by a series of $Z + 1$ nested hypotheses each consisting of two inequality constrained components. Addition of these $L \times 2 \times (Z + 1)$ constraints to R_m renders a constraints matrix with $Q' = Q + L \times 2 \times (Z + 1)$ rows that can be decomposed as in (10.50). It is straightforward to prove (see Section 10.2) that

$$BF_{m_{Q'-v}, m_{Q'-v-2}} = \frac{f_{m_{Q'-v}, m_{Q'-v-2}}}{c_{m_{Q'-v}, m_{Q'-v-2}}}, \tag{10.51}$$

where $f_{m_{Q'-v}, m_{Q'-v-2}}$ denotes the proportion of the posterior distribution of $H_{m_{Q'-v-2}}$ in agreement with $H_{m_{Q'-v}}$, and $c_{m_{Q'-v}, m_{Q'-v-2}}$ the proportion of the prior distribution of $H_{m_{Q'-v-2}}$ in agreement with $H_{m_{Q'-v}}$.

Using constrained sampling, which will be elaborated in the next section, each element of (10.50) can accurately be computed using a relatively small sample from the corresponding prior and posterior distributions. Note again that the simple strategy needs a sample of more than 20 million in order to obtain an accurate estimate of BF_{ma} for $H_m : \mu_1 > \ldots > \mu_{10}$. Use of (10.50) leads to a decomposition of BF_{ma} into five Bayes factors. Because

$$
\boldsymbol{R}_m = \begin{bmatrix}
1 & -1 & 0 & 0 & 0 & 0 & 0 & 0 & 0 & 0 \\
0 & 1 & -1 & 0 & 0 & 0 & 0 & 0 & 0 & 0 \\
0 & 0 & 1 & -1 & 0 & 0 & 0 & 0 & 0 & 0 \\
0 & 0 & 0 & 1 & -1 & 0 & 0 & 0 & 0 & 0 \\
0 & 0 & 0 & 0 & 1 & -1 & 0 & 0 & 0 & 0 \\
0 & 0 & 0 & 0 & 0 & 1 & -1 & 0 & 0 & 0 \\
0 & 0 & 0 & 0 & 0 & 0 & 1 & -1 & 0 & 0 \\
0 & 0 & 0 & 0 & 0 & 0 & 0 & 1 & -1 & 0 \\
0 & 0 & 0 & 0 & 0 & 0 & 0 & 0 & 1 & -1
\end{bmatrix}, \tag{10.52}
$$

these Bayes factors are BF_{2a}, that is, a constrained hypothesis build using the first two constraints in (10.52) evaluated against the unconstrained hypothesis; BF_{42}; BF_{64}; BF_{86}; and BF_{98}. The corresponding complexity values are $c_{2a} = 1/6$; $c_{42} = 1/20$; $c_{64} = 1/42$; $c_{86} = 1/72$; and $c_{98} = 1/10$. Each of these complexity values can accurately be estimated using a sample of, say, 20,000 from the corresponding prior distributions (which is substantially less than a sample of more than 20 million and nevertheless renders an increased accuracy). Note that the multiplication of the five complexity values is exactly equal to $1/10!$.

10.5.2 Constrained Sampling

Sampling from the prior and posterior distribution of $H_{m_{Q'-v}}$, where $Q' = Q + L \times 2 \times (Z+1)$ and $v = Q', \ldots, 4, 2, 0$ (note that $H_a = H_{m_0}$) can for all statistical models discussed in this book be achieved using a Markov chain Monte Carlo method (Robert and Casella, 2004) like the Gibbs sampler (Casella and George, 1992). Note that for the multilevel model and the latent class model that were discussed in Sections 7.3 and 7.4, respectively, a data-augmented Gibbs sampler (Zeger and Karim, 1991) is needed. The interested reader is referred to Klugkist and Mulder (2008) for an accessible account of constrained sampling in the context of constrained analysis of variance models.

The general idea of the Gibbs sampler is to sequentially sample each (group of) parameter(s) from the prior or posterior distribution conditionally on the current values of the other parameters. Let, as before, $\boldsymbol{\theta}$ denote the parameters subjected to constraints and $\boldsymbol{\phi}$ the nuisance parameters. Let $\boldsymbol{\xi}$ denote latent parameters (like a person's class membership in latent class analysis) or random parameters (like the random intercepts and slopes in multilevel analysis). In its data-augmented form, the Gibbs sampler has the following form:

- Provide initial values for $\boldsymbol{\theta}^0$ that are in agreement with the constraints in $H_{m_{Q'-v}}$ and provide initial values for $\boldsymbol{\phi}^0$.

Repeat the following steps $t = 1, \ldots, T$ times:

- Sample $\boldsymbol{\xi}^t$ conditional on $\boldsymbol{\theta}^{t-1}$ and $\boldsymbol{\phi}^{t-1}$.

- Sample $\boldsymbol{\phi}^t$ conditional on the current values of $\boldsymbol{\theta}^{t-1}$ and $\boldsymbol{\xi}^t$.

- Do for $d = 1, \ldots, D$: sample θ_d^t from its distribution conditional on $\boldsymbol{\xi}^t$, $\boldsymbol{\phi}^t$ and the current values of the other θs. In combination with the $Q' - v$ constraints that are currently active, the current values of the other θs can be used to determine a lower bound L and upper bound U for θ_d^t. Using inverse probability sampling (Gelfand, Smith, and Lee, 1992) it is straightforward to obtain a sample from a distribution with lower bound L and upper bound U:

- Sample a random number ν from a uniform distribution on the interval $[0, 1]$.
- Compute the proportions p_L and p_U of the posterior distribution of θ_d that are not admissible due to L and U:

$$p_L = \int_{nL}^{L} g(\theta_d \mid \phi^t, \xi^t, L, B, Y, X, H_{m_{Q'-v}}) \partial \theta_d, \tag{10.53}$$

and

$$p_U = \int_{U}^{nU} g(\theta_d \mid \phi^t, \xi^t, L, B, Y, X, H_{m_{Q'-v}}) \partial \theta_d, \tag{10.54}$$

where nL and uL denote the natural lower and upper bounds of the distribution at hand. For example, for a normal distribution, these are $-\infty$ and ∞, respectively; and for a beta distribution, 0 and 1, respectively. Note that for the prior distribution analogous formulas can be derived.

- Compute θ_d^t such that it is the deviate associated with the ν-th percentile of the admissible part of the posterior distribution of θ_d:

$$p_L + \nu(1 - p_L - p_U) = \int_{nL}^{\theta_d^t} g(\theta_d \mid \phi^t, \xi^t, L, B, Y, X, H_{m_{Q'-v}}) \partial \theta_d. \tag{10.55}$$

There are many examples of the procedure sketched in this section. See, for example, Klugkist, Laudy, and Hoijtink (2005) in the context of ANCOVA models; Mulder, Hoijtink, and Klugkist (2010) in the context of the multivariate normal linear model; Hoijtink (1998) in the context of the latent class model; Kato and Hoijtink (2006) in the context of multilevel models; and Laudy and Hoijtink (2007) in the context of contingency tables. The procedure is fully implemented in the software packages BIEMS (see Section 9.2.2) and ContingencyTable (see Section 9.2.5). In ConfirmatoryANOVA (see Section 9.2.1), it is only implemented for means restricted to be equal to each other.

10.5.3   Unidentified Models

As long as the statistical model of interest is identified, the computation of $f_{m_{Q'-v}, m_{Q'-v-2}}$ and $c_{m_{Q'-v}, m_{Q'-v-2}}$ is straightforward: determine the proportion of parameter vectors sampled from the posterior and prior distribution of $H_{m_{Q'-v-2}}$ in agreement with $H_{m_{Q'-v}}$. Note that a model is unidentified if for each combination of parameter values there exist one or more equivalent representations, that is, representations that render the same density of the data. With the exception of the latent class model (see Section 7.4) all the models discussed in this book are identified. A sample from the prior or posterior distribution of the latent class model renders vectors θ_d, ϕ_d for $d = 1, \ldots, D$ latent classes. The latent class model is not identified in the sense that the D class labels may be permuted (this phenomenon is known as label switching; see Stevens (2000)). This immediately implies that the computation of $f_{m_{Q'-v}, m_{Q'-v-2}}$ and $c_{m_{Q'-v}, m_{Q'-v-2}}$ is less straightforward than before: determine the proportion of parameter vectors sampled from the posterior and prior distribution of $H_{m_{Q'-v-2}}$ for which at least one permutation is in agreement with $H_{m_{Q'-v}}$. The general rule underlying this exposition is formulated in the following definition:

Definition 10.4: If a model is not identified, complexity and fit can be determined via the computation of the proportion of parameter vectors (sampled from prior and posterior distributions, respectively) that are themselves or in one of their equivalent representations in agreement with the hypothesis being evaluated.

10.5.4   Monte Carlo Uncertainty

As was elaborated in Section 10.5.2, each element of (10.50) can be estimated using estimates of $f_{m_{Q'-v},m_{Q'-v-2}} = f_v$ and $c_{m_{Q'-v},m_{Q'-v-2}} = c_v$ based on a sample of size T from the corresponding posterior and prior distribution of $\boldsymbol{\theta}$. For $T \to \infty$, the variance of the estimator of each element and consequently of (10.50) goes to 0. For finite T, an estimate of the variance of (10.50) due to sampling, the so-called Monte Carlo error, can be obtained using simulation.

Both c_v and f_v are parameters in a binomial process that renders T_{c_v} and T_{f_v} vectors $\boldsymbol{\theta}$, respectively, that are in agreement with $H_{m_{Q'-v}}$. Using a uniform Beta prior distribution $\mathcal{B}(1,1)$ for both c_v and f_v their posterior distributions are $\mathcal{B}(T_{c_v} + 1, T - T_{c_v} + 1)$ and $\mathcal{B}(T_{f_v} + 1, T - T_{f_v} + 1)$, respectively.

An estimate of the 95% central credibility interval of (10.50) can be obtained as follows:

- Sample c_{vu} for $u = 1, \ldots, U$ from $\mathcal{B}(T_{c_v} + 1, T - T_{c_v} + 1)$ for all v.

- Sample f_{vu} for $u = 1, \ldots, U$ from $\mathcal{B}(T_{f_v} + 1, T - T_{f_v} + 1)$ for all v.

- For $u = 1, \ldots, U$ compute $BF_{ma,u}$ using (10.50).

- Compute the 95% central credibility interval for BF_{ma} using the 2.5-th and 97.5-th percentile of the U sampled values of $BF_{ma,u}$, or, compute the standard deviation of BF_{ma}, that is, compute the standard deviation of the U sampled values of $BF_{ma,u}$.

The simplicity of this procedure ensures that it can be executed rather quickly, even for large values of U. As illustrated in Appendices 7C en 7F, WinBUGS can be used to compute a 95% central credibility interval for BF_{ma}. The software package BIEMS renders the standard deviation of BF_{ma}.

10.6   Discussion

In this chapter the foundations of Bayesian evaluation of informative hypotheses were presented. First of all, it was shown that the Bayes factor has a simple form based on prior and posterior proportions if the goal is to evaluate informative hypotheses. Subsequently the requirement for the specification of the prior distribution and the properties of Bayes factors based on the prior distributions were elaborated for the multivariate normal linear model and statistical models in general. After a short section dealing with the posterior distribution, estimation of the prior and posterior proportions needed in order to be able to compute the Bayes factor was discussed. Among other things, a decomposition of the Bayes factor, augmented Gibbs sampling, constrained sampling, unidentified models, and Monte Carlo errors were discussed.

10.7   Appendix 10A: Density of the Data of Various Statistical Models

Univariate Normal Linear Model (Chapter 1)

$$f(\boldsymbol{y} \mid \boldsymbol{X}, \boldsymbol{\mu}, \boldsymbol{\beta}, \sigma^2) =$$

$$\prod_{i=1}^{N} \frac{1}{\sqrt{2\pi\sigma^2}} \exp \frac{(y_i - \mu_1 d_{1i} - \ldots - \mu_J d_{Ji} - \beta_1 x_{1i} - \ldots - \beta_K x_{Ki})^2}{-2\sigma^2},$$

where $\boldsymbol{y} = [y_1, \ldots, y_N]$, $\boldsymbol{X} = [\boldsymbol{d}_1, \boldsymbol{x}_1, \ldots, \boldsymbol{d}_N, \boldsymbol{x}_N]$, $\boldsymbol{d}_i = [d_{1i}, \ldots, d_{Ji}]$, where $d_{ji} = 1/0$ denotes that person i is/is not a member of group j, $\boldsymbol{x}_i = [x_{1i}, \ldots, x_{Ki}]$, $\boldsymbol{\mu} = [\mu_1, \ldots, \mu_J]$, and $\boldsymbol{\beta} = [\beta_1, \ldots, \beta_K]$.

Multivariate Normal Linear Model (Chapter 2)

$$f(\boldsymbol{Y} \mid \boldsymbol{X}, \boldsymbol{\mu}, \boldsymbol{\beta}, \boldsymbol{\Sigma}) = \prod_{i=1}^{N} \frac{1}{2\pi^{P/2}|\boldsymbol{\Sigma}|^{1/2}} \exp -\frac{1}{2} \boldsymbol{\epsilon}_i \boldsymbol{\Sigma}^{-1} \boldsymbol{\epsilon}_i,$$

where $\boldsymbol{Y} = [\boldsymbol{y}_1, \ldots, \boldsymbol{y}_N]$, with $\boldsymbol{y}_i$ a vector of length P, $\boldsymbol{\mu}$ a $P \times J$ matrix with elements μ_{pj}, $\boldsymbol{\beta}$ a $P \times K$ matrix with elements β_{pk}, and $\boldsymbol{\epsilon}_i = \boldsymbol{y}_i - \boldsymbol{\mu} \boldsymbol{d}_i - \boldsymbol{\beta} \boldsymbol{x}_i$.

Binomial Distribution (Chapter 10)

$$f(y \mid N, \pi) = \binom{N}{y} \pi^y (1 - \pi)^{N-y},$$

where y denotes the number of successes in N trials, and π the probability of success.

Contingency Tables (Chapter 7)

$$f(\boldsymbol{y} \mid \boldsymbol{\pi}) = \binom{N}{\prod_{d-1}^{D} y_d} \prod_{d=1}^{D} \pi_d^{y_d},$$

where $\boldsymbol{y} = [y_1, \ldots, y_D]$, y_d is the frequency observed in cell d of the contingency table, and π_d the corresponding probability. Note that, $\boldsymbol{\pi} = [\pi_1 \ldots, \pi_D]$.

Multilevel Model (Chapter 7)

$$f(\boldsymbol{y} \mid \boldsymbol{x}, \boldsymbol{\beta}, \boldsymbol{\mu}, \boldsymbol{\phi}, \sigma^2) = \prod_{i=1}^{N} \int_{\boldsymbol{\beta}_i} \prod_{t=1}^{4} \frac{1}{\sqrt{2\pi\sigma^2}} \exp \frac{(y_{it} - \beta_{0i} - \beta_{1i} x_t)^2}{-2\sigma^2} \mathcal{N}(\boldsymbol{\beta}_i \mid \boldsymbol{\mu}, \boldsymbol{\phi}) \partial \boldsymbol{\beta}_i,$$

where $\boldsymbol{y} = [\boldsymbol{y}_1, \ldots, \boldsymbol{y}_N]$ with $\boldsymbol{y}_1 = [y_{i1}, \ldots, y_{i4}]$, $\boldsymbol{\beta} = [\boldsymbol{\beta}_1, \ldots, \boldsymbol{\beta}_N]$ with $\boldsymbol{\beta}_i = [\beta_{0i}, \beta_{1i}]$, $\boldsymbol{\mu} = [\mu_0, \mu_1]$, $\boldsymbol{\phi}$ is a 2×2 covariance matrix with ϕ_{00} and ϕ_{11} on the diagonal and ϕ_{01} off the diagonal, and $\boldsymbol{x} = [x_1, \ldots, x_4] = [-1.5, -.5, .5, 1.5]$.

Latent Class Model (Chapter 7)

$$f(\boldsymbol{Y} \mid \boldsymbol{\pi}, \boldsymbol{\omega}) = \prod_{i=1}^{N} \sum_{d=1}^{D} \omega_d \prod_{j=1}^{J} \pi_{dj}^{y_{ij}} (1 - \pi_{dj})^{(1-y_{ij})},$$

where $\boldsymbol{Y}$ is a $N \times J$ matrix with entries $y_{ij} \in \{0, 1\}$ denoting the response of person i to question j, $\boldsymbol{\pi}$ is a $D \times J$ matrix with entries π_{dj}, and $\boldsymbol{\omega} = [\omega_1, \ldots, \omega_D]$.

10.8 Appendix 10B: Unconstrained Prior Distributions Used in Book and Software

Univariate Normal Linear Model (Chapter 1)

$$h(\boldsymbol{\theta} \mid H_a)h(\sigma^2) = \mathcal{N}(\boldsymbol{\mu}_0^r, \boldsymbol{T}_0^r)\text{Inv-}\chi^2(\nu_0, \sigma_0^2).$$

Note that in BIEMS, $h(\sigma^2) \propto 1/\sigma^2$.

Multivariate Normal Linear Model (Chapter 2)

$$h(\boldsymbol{\theta} \mid H_a)h(\boldsymbol{\Sigma}) = \mathcal{N}(\boldsymbol{\mu}_0^r, \boldsymbol{T}_0^r)\text{Inv-}\mathcal{W}(\nu_0, \boldsymbol{\Sigma}_0).$$

Note that in BIEMS, $h(\boldsymbol{\Sigma}) \propto |\boldsymbol{\Sigma}|^{-(P+1)/2}$.

Binomial Distribution (Chapter 10)

$$h(\pi \mid H_a) = \mathcal{B}(a_0, b_0).$$

Contingency Tables (Chapter 7)

$$h(\pi_1 \ldots, \pi_D \mid H_a) = \mathcal{D}(a_0, \ldots, a_0).$$

Multilevel Model (Chapter 7)

$$h(\mu_0, \mu_1 \mid H_a)h(\boldsymbol{\phi})h(\sigma^2) = \mathcal{N}(\boldsymbol{\mu}_0^r, \boldsymbol{T}_0^r)\text{Inv-}\mathcal{W}(\nu_0, \boldsymbol{\Sigma}_0)\text{Inv-}\chi^2(\nu_0, \sigma_0^2).$$

Note that in WinBUGS (see Section 9.2.7) the prior distribution of the parameters of the multilevel model is parameterized as

$$h(\mu_0, \mu_1 \mid H_a)h(\boldsymbol{\phi}^{-1})h(\sigma^{-2}) = \mathcal{N}(\boldsymbol{\mu}_0^r, \boldsymbol{T}_0^r)\mathcal{W}(\nu_0, \boldsymbol{\Sigma}_0)\mathcal{G}(a, b).$$

Note that the parametrization of the Wishart distribution in WinBUGS is not the standard parametrization. This is elaborated in Appendix 10C.

Latent Class Model (Chapter 7)

$$h(\boldsymbol{\pi} \mid H_a) \times h(\boldsymbol{\omega}) = \prod_{d=1}^{D} \prod_{j=1}^{J} \mathcal{B}(a_0, b_0) \times \mathcal{D}(a_0, \ldots, a_0).$$

10.9   Appendix 10C: Probability Distributions Used in Appendices 10A and 10B

The Multivariate Normal Distribution

$$\mathcal{N}(\boldsymbol{\mu}_0^r, \boldsymbol{T}_0^r) = \frac{1}{2\pi^{D/2}|\boldsymbol{T}_0^r|^{1/2}} \exp{-\frac{1}{2}(\boldsymbol{\theta} - \boldsymbol{\mu}_0^r)(\boldsymbol{T}_0^r)^{-1}(\boldsymbol{\theta} - \boldsymbol{\mu}_0^r)}.$$

The Scaled Inverse Chi-Square Distribution

$$\text{Inv-}\chi^2(\nu_0, \sigma_0^2) \propto \frac{\exp{\frac{-\nu_0\sigma_0^2}{2\sigma^2}}}{(\sigma^2)^{1+\nu_0/2}}.$$

Using $\nu_0 = 0$ and an arbitrary value for σ_0^2 renders Jeffreys prior $1/\sigma^2$ as implemented in BIEMS. Using $\nu_0 = 1$ and $\sigma_0^2 = s^2$, where s^2 is an estimate of the residual variance, renders a vague prior distribution. This prior distribution is implemented in `ConfirmatoryANOVA`.

The Inverse Wishart Distribution

$$\text{Inv-}\mathcal{W}(\nu_0, \boldsymbol{\Sigma}_0) \propto$$

$$|\boldsymbol{\Sigma}|^{-(\nu_0+P+1)/2} \exp(-\frac{1}{2}\text{tr}(\boldsymbol{\Sigma}_0\boldsymbol{\Sigma}^{-1})).$$

Using $\nu_0 = 0$ and $\boldsymbol{\Sigma}_0 = \boldsymbol{0}$ renders Jeffreys prior $|\boldsymbol{\Sigma}|^{-(P+1)/2}$ as implemented in BIEMS. Noting that $E(\boldsymbol{\Sigma}) = \boldsymbol{\Sigma}_0/(\nu_0 - P - 1)$, a vague specification can be obtained using $\nu_0 = P+2$ and $\boldsymbol{\Sigma}_0 = \hat{\boldsymbol{\Sigma}}$, where $\hat{\boldsymbol{\Sigma}}$ denotes an estimate of the residual covariance matrix obtained under H_a. The result is that a priori it is specified that $E(\boldsymbol{\Sigma}) = \hat{\boldsymbol{\Sigma}}$ with low certainty, that is, $\nu_0 = P+2$.

The Wishart Distribution

$$\mathcal{W}(\nu_0, \boldsymbol{\Sigma}_0) \propto$$

$$|\boldsymbol{\Sigma}^{-1}|^{(\nu_0-P-1)/2} \exp(-\frac{1}{2}\text{tr}(\boldsymbol{\Sigma}_0^{-1}\boldsymbol{\Sigma}^{-1})).$$

The Wishart distribution is used in `WinBUGS` to specify a prior distribution for the inverse covariance matrix $\boldsymbol{\Sigma}^{-1}$. Noting that $E(\boldsymbol{\Sigma}^{-1}) = \nu_0\boldsymbol{\Sigma}_0$, a vague specification can be obtained using $\nu_0 = P$ and $\boldsymbol{\Sigma}_0 = \hat{\boldsymbol{\Sigma}}^{-1}/P$, where $\hat{\boldsymbol{\Sigma}}$ denotes an estimate of the residual covariance matrix obtained under H_a. The result is that a priori it is specified that $E(\boldsymbol{\Sigma}^{-1}) = \hat{\boldsymbol{\Sigma}}^{-1}$ with low certainty, that is, $\nu_0 = P$. Note that `WinBUGS` uses a nonstandard parametrization of the Wishart distribution. In the `WinBUGS` parametrization, the Wishart distribution described here is obtained if the `WinBUGS`-$\boldsymbol{\Sigma}_0$ is set equal to $\hat{\boldsymbol{\Sigma}} \times P$.

The Beta Distribution

$$\mathcal{B}(a_0, b_0) = \frac{\Gamma(a_0 + b_0)}{\Gamma(a_0)\Gamma(b_0)} \pi^{a_0-1}(1 - \pi)^{b_0-1}.$$

Using $a_0 = b_0 = 1$, a uniform prior distribution on the interval [0,1] is obtained. This prior distribution is implemented in the software package `ContingencyTable` and was used for the class-specific probabilities in the `WinBUGS` application presented in Appendix 7D.

The Dirichlet Distribution

$$\mathcal{D}(a_0, \ldots, a_0) = \frac{\Gamma(\sum_{d=1}^{D} a_0)}{\prod_{d=1}^{D} \Gamma(a_0)} \prod_{d=1}^{D} \pi_d^{a_0 - 1}.$$

Using $a_0 = 1$, a standard noninformative distribution is obtained in which all combinations of parameters values are a priori equally likely. This prior distribution was used for the class weights in the `WinBUGS` application presented in Appendix 7E. It is also the prior distribution implemented in `ContingencyTable` for the cell probabilities of a contingency table (see Sections 7.2 and 9.2.5).

The Gamma Distribution

$$\mathcal{G}(a, b) = \frac{b^a}{\Gamma(a)} (\sigma^{-2})^{a-1} \exp -b\sigma^{-2}.$$

The Gamma distribution is used in `WinBUGS` to specify a prior distribution for the inverse variance σ^{-2}. Using small values for the parameters, say $a = .01$ and $b = .01$, a vague distribution is specified.

References

Agresti, A. (2007). *An Introduction to Categorical Data Analysis*. New York: John Wiley & Sons.

Akaike, H. (1987). Factor analysis and AIC. *Psychometrika, 52*, 317–332.

Allen, M.P. (1997). *Understanding Regression Analysis*. New York: Springer.

Allison, P.D. (1999). *Multiple Regression. A Primer*. Thousand Oaks, CA: Pine Forge Press.

Anraku, K. (1999). An information criterion for parameters under a simple order restriction. *Biometrika, 86*, 141–152.

Barker Bausell, R. and Li, Y.-F. (2006). *Power Analysis for Experimental Design. A Practical Guide for the Biological, Medical and Social Sciences*. Cambridge: Cambridge University Press.

Barlow, R.E., Bartholomew, D.J., Bremner, H.M., and Brunk, H.D. (1972). *Statistical Inference under Order Restrictions*. New York: Wiley.

Barnett, V. (1999). *Comparative Statistical Inference*. New York: John Wiley & Sons.

Bayarri, M.J. and Berger, J.O. (2004). The interplay of Bayesian and frequentist analysis. *Statistical Science, 19*, 58–80.

Berger, J.O. (2003). Could Fisher, Jeffreys and Neyman have agreed on testing? *Statistical Science, 18*, 1–32.

Berger, J.O., Brown, L.D., and Wolpert, R.L. (1994). A unified conditional frequentist and Bayesian test for fixed and sequential simple hypothesis testing. *The Annals of Statistics, 4*, 1787–1807.

Berger, J.O. and Delampady, M. (1987). Testing precise hypotheses. *Statistical Science, 2*, 317–352.

Berger, J.O. and Mortera, J. (1999). Default Bayes factors for non-nested hypothesis testing. *Journal of the American Statistical Association, 94*, 542–554.

Berger, J.O. and Perricchi, L. (1996). The intrinsic Bayes factor for model selection and prediction. *Journal of the American Statistical Association, 91*, 109–122.

Berger, J.O. and Perricchi, L. (2004). Training samples in objective Bayesian model selection. *Annals of Statistics, 32*, 841–869.

Berger, J.O. and Sellke, T. (1987). Testing a point null hypothesis: The irreconcilability of *P* values and evidence. *Journal of the American Statistical Association, 82*, 112–122.

Bullens, J., Klugkist, I., and Postma, A. (In Press). The role of local and distal landmarks in the development of object location memory. *Developmental Psychology.*

Burnham, K.P. and Anderson, D.R. (2002). *Model Selection and Multi Model Inference. A Practical Information-Theoretic Approach.* New York: Springer.

Carlin, B.P. and Chib, S. (1995). Bayesian model choice via Markov chain Monte Carlo methods. *Journal of the Royal Statistical Society, Series B, 57,* 473–484.

Carnap, R. (1950). *Logical Foundations of Probability.* Chicago: University of Chicago Press.

Carnap, R. (1952). *The Continuum of Inductive Methods.* Chicago: University of Chicago Press.

Casella, G. and Berger, R.L. (1987). Reconciling Bayesian and frequentist evidence in the one-sided testing problem. *Journal of the American Statistical Association, 82,* 106–111.

Casella, G. and George, E.I. (1992). Explaining the Gibbs sampler. *The American Statistician, 46,* 167–174.

Chib, S. (1995). Marginal likelihood from the Gibbs output. *Journal of the American Statistical Association, 90,* 1313–1321.

Cohen, J. (1968). Multiple regression as a general data analytic system. *Psychological Bulletin, 70,* 426–443.

Cohen, J. (1988). *Statistical Power Analysis for the Behavioral Sciences.* Hilldale, NJ: Lawrence Erlbaum Associates.

Cohen, J. (1992). A power primer. *Psychological Bulletin, 112,* 155–159.

Cohen, J. (1994). The earth is round, p<.05. *American Psychologist, 49,* 997–1003.

Congdon, P. (2003). *Applied Bayesian Modelling.* New York: Wiley.

Croon, M. A. (1990). Latent class analysis with ordered latent classes. *British Journal of Mathematical and Statistical Psychology, 43,* 171–192.

Croon, M. A. (1991). Investigating Mokken scalability of dichotomous items by means of ordinal latent class analysis. *British Journal of Mathematical and Statistical Psychology, 44,* 315–331.

Davis, C.S. (2002). *Statistical Methods for the Analysis of Repeated Measurements.* New York: Springer.

De Leeuw, E.D., Hox, J.J., and Dillman, D.A. (2008). *International Handbook of Survey Methodology.* New York: Lawrence Erlbaum Associates, Taylor and Francis Group.

De Leeuw, C. and Mulder, J. (Unpublished). Manual for BIEMS Data Generator.

De Santis, F. (2004). Statistical evidence and sample size determination for Bayesian hypothesis testing. *Journal of Statistical Planning and Inference, 124,* 121–144.

De Vaus, D.A. (2001). *Research Design in Social Research.* London: SAGE.

De Vaus, D.A. (2002). *Surveys in Social Research.* Crows Nest, Australia: Allen and Unwin.

Dickey, J. (1971). The weighted likelihood ratio, linear hypotheses on normal location parameters. *The Annals of Statistics, 42,* 204-223.

Garcia-Donato, G. and Chen, M.-H. (2005). Calibrating Bayes factor under prior predictive distributions. *Statistics Sinica, 15,* 359–380.

Gelfand, A.E., Smith, A.F.M., and Lee, T. (1992). Bayesian analysis of constrained parameter and truncated data problems using Gibbs sampling. *Journal of the American Statistical Association, 87,* 523–532.

Gelman, A., Carlin, J.B., Stern, H.S., and Rubin, D.B. (2004). *Bayesian Data Analysis.* Boca Raton, FL: Chapman & Hall/CRC.

Gill, J. (2002). *Bayesian Methods. A Social and Behavioral Sciences Approach.* Boca Raton, FL: Chapman & Hall/CRC.

Han, C. and Carlin, P. (2001). Markov chain Monte Carlo methods for computing Bayes factors: A comparative review. *Journal of the American Statistical Association, 96,* 1122–1132.

Hoijtink, H. (1998). Constrained latent class analysis using the Gibbs sampler and posterior predictive p-values: Applications to educational testing. *Statistica Sinica, 8,* 691–712.

Hoijtink, H. (2001). Confirmatory latent class analysis: Model selection using Bayes factors and (pseudo) likelihood ratio statistics. *Multivariate Behavioral Research, 36,* 563–588.

Hoijtink, H. (2009). Bayesian data analysis. In R.E. Millsap and A. Maydeu-Olivares (Eds), *The SAGE Handbook of Quantitave Methods in Psychology,* 423–443. London: SAGE.

Hoijtink, H. (Unpublished). Objective Bayes Factors for Inequality Constrained Hypotheses.

Hoijtink, H. and Boom, J. (2008). Inequality constrained latent class analysis. In: H. Hoijtink, I. Klugkist and P.A. Boelen (Eds.), *Bayesian Evaluation of Informative Hypotheses,* pp. 227–246. New York: Springer.

Hoijtink, H., Klugkist, I., and Boelen, P.A. (2008). *Bayesian Evaluation of Informative Hypotheses.* New York: Springer.

Hoijtink, H. and Molenaar, I.W. (1997). A multidimensional item response model: Constrained latent class analysis using the Gibbs sampler and posterior predictive checks. *Psychometrika, 62,* 171–190.

Hosmer, D.W. and Lemeshow, S. (2000). *Applied Logistic Regression.* New York: John Wiley & Sons.

Howard, G.S., Maxwell, S.E., and Fleming, K. (2000). The proof of the pudding: An illustration of the relative strengths of null hypothesis, meta-analysis, and Bayesian analysis. *Psychological Methods, 5,* 315–232.

Howell, D.C. (2009). *Statistical Methods for Psychology.* Florence, KY: Cengage Learning.

Howson, C. and Urbach, P. (2006). *Scientific Reasoning. The Bayesian Approach.* Peru, IL: Open Court Publishing Company.

Hox, J.J. (2010). *Multilevel Analysis. Techniques and Applications.* New York: Routledge.

Huber, P.J. (1981). *Robust Statistics.* New York: John Wiley & Sons.

Jaynes, E.T. (2003). *Probability Theory. The Logic of Science.* Cambridge: Cambridge University Press.

Jeffreys, H. (1961). *Theory of Probability. Third Edition.* Oxford: Oxford University Press.

Kammers, M., Mulder, J., de Vignemont, F., and Dijkerman, H.C. (2009). The weight of representing the body: Addressing the potentially indefinite number of body representations in healthy individuals. *Experimental Brain Research, 204,* 333–342.

Kass, R.E. and Raftery, A.E. (1995). Bayes factors. *Journal of the American Statistical Association, 90,* 773–795.

Kass, R.E. and Wasserman, L. (1996). The selection of prior distributions by formal rules. *Journal of the American Statistical Association, 91,* 1343–1370.

Kato, B.S. and Hoijtink, H. (2006). A Bayesian approach to inequality constrained linear mixed models: Estimation and model selection. *Statistical Modelling, 6,* 231–249.

Kato, B.S. and Peeters, C. (2008). Inequality constrained multilevel models. In: H. Hoijtink, I. Klugkist and P.A. Boelen (Eds.), *Bayesian Evaluation of Informative Hypotheses,* pp. 273–298. New York: Springer.

Kim, K. and Timm, N. (2007). *Univariate and Multivariate General Linear Models. Theory and Applications with SAS.* New York: Chapmann & Hall/CRC.

Kleinbaum, D.G., Kupper, L.L., and Muller, K.E. (2008). *Applied Regression Analysis and other Multivariate Methods.* Florence, KY: Cengage Learning.

Kline, R.B. (2005). *Principles and Practice of Structural Equation Modeling.* New York: Guiford Press.

Klugkist, I. and Hoijtink, H. (2007). The Bayes factor for inequality and about equality constrained models. *Computational Statistics and Data Analysis, 51,* 6367–6379.

Klugkist, I., Kato, B., and Hoijtink, H. (2005). Bayesian model selection using encompassing priors. *Statistica Neerlandica, 59,* 57–69.

Klugkist, I., Laudy, O., and Hoijtink, H. (2005). Inequality constrained analysis of variance: A Bayesian approach. *Psychological Methods, 10,* 477–493.

Klugkist, I., Laudy, O., and Hoijtink, H. (2010). Bayesian evaluation of inequality and equality constrained hypotheses for contingency tables. *Psychological Methods, 15,* 281–299.

Klugkist, I. and Mulder, J. (2008). Bayesian estimation of inequality constrained analysis of variance. In: H. Hoijtink, I. Klugkist, and P.A. Boelen (Eds.), *Bayesian Evaluation of Informative Hypotheses,* pp. 27–52. New York: Springer.

Kuiper, R.M. and Hoijtink, H. (2010). Comparisons of means using exploratory and confirmatory approaches. *Psychological Methods, 15,* 69–86.

Kuiper, R.M. and Hoijtink, H. (Unpublished). A Fortran 90 program for the generalization of the order restricted information criterion.

Kuiper, R.M., Hoijtink, H., and Silvapulle, M.J. (2011). An Akaike type information criterion for model selection under inequality constraints. *Biometrika, 98,* 495–501.

Kuiper, R.M., Hoijtink, H., and Silvapulle, M.J. (Unpublished). Generalization of the order restricted information criterion for multivariate normal linear models.

Kuiper, R. M., Klugkist, I., and Hoijtink, H. (2010). A Fortran 90 program for confirmatory analysis of variance. *Journal of Statistical Software, 34*, 1–31.

Kuiper, R.M., Nederhoff, T., and Klugkist, I. (Unpublished). Performance of exploratory and confirmatory approaches.

Laudy, O. (2008). Inequality constrained contingency table analysis. In: H. Hoijtink, I. Klugkist, and P.A. Boelen (Eds.), *Bayesian Evaluation of Informative Hypotheses*, pp. 247–272. New York: Springer.

Laudy, O., Boom, J., and Hoijtink, H. (2004). Bayesian computational methods for inequality constrained latent class analysis. In: A. van der Ark, M. Croon, and K. Sijtsma (Eds.), *New Developments in Categorical Data Analysis for the Social and Behavioral Sciences*, pp. 63–82. Mahwah, NJ: Erlbaum.

Laudy, O. and Hoijtink, H. (2007). Bayesian methods for the analysis of inequality constrained contingency tables. *Statistical Methods in Medical Research, 16*, 123–138.

Laudy, O., Zoccolillo, M., Baillargeon, R., Boom, J., Tremblay, R., and Hoijtink, H. (2005). Applications of confirmatory latent class analysis in developmental psychology. *European Journal of Developmental Psychology, 2*, 1–15.

Lavine, M. and Schervish, M.J. (1999). Bayes factors: What they are and what they are not. *The American Statistician, 53*, 119–122

Leucari, V. and Consonni, G. (2003). Compatible priors for causal Bayesian networks. In: J.M. Bernardo, M.J. Bayarri, J.O. Berger, A.P. Dawid, D. Heckerman, A.F.M. Smith, and M.West, *Bayesian Statistics 7*, pp. 597–606. Oxford: Clarendon Press.

Li, G., Gao, X., and Huang, M. (2003). Testing multivariate one-sided hypotheses. *Statistics & Probability Letters, 64*, 63–68.

Lindley, D.V. (1957). A statistical paradox. *Biometrika, 44*, 187–192.

Lucas, J.W. (2003). Status processes and the institutionalization of women as leaders. *American Sociological Review, 68*, 464–480.

Lynch, S.M. (2007). *Introduction to Applied Bayesian Statistics and Estimation for Social Scientists*. New York: Springer.

Maxwell, S.E. (2004). The persistence of underpowered studies in psychological research, causes, consequences and remedies. *Psychological Methods, 9*, 147–163.

McCullagh, P. and Nelder, J. (1989). *Generalized Linear Models, Second Edition*. Boca Raton, FL: Chapman and Hall/CRC

McCullogh, C.E. and Searle, S.R. (2001). *Generalized Linear and Mixed Models*. New York: Wiley.

McCutcheon, A.L. (1987). *Latent Class Analysis*. Thousand Oaks, CA: SAGE.

Meeus, W., van de Schoot, R., Keijsers, L., Schwartz, S.J., and Branje, S. (2010). On the progression and stability of adolescent identity formation. A five-wave longitudinal study in early-to-middle and middle-to-late adolescence. *Child Development, 81*, 1565–1581.

Meeus, W., van de Schoot, R., Klimstra, T., and Branje, S. (2011). Change and stability of personality types in adolescence: A five-wave longitudinal study in early-to-middle and middle-to-late adolescence. *Developmental Psychology, 47,* 1181–1195.

Miller, R.G. (1997). *Beyond ANOVA. Basics of Applied Statistics.* Boca Raton, FL: Chapman & Hall/CRC.

Molenberghs, G. and Verbeke, G. (2007). Likelihood ratio, score, and wald tests in a constrained parameter space. *The American Statistician, 61,* 22–27.

Moreno, E. (2005). Objective Bayesian methods for one-sided testing. *Test, 14,* 181–198.

Mulder, J. (Unpublished). A paradox with balances Bayes factors.

Mulder, J. and Hoijtink, H. (Unpublished). Default Bayes factors for type B testing problems.

Mulder, J., Hoijtink, H., and de Leeuw, C. (In Press). BIEMS: A Fortran90 program for calculating Bayes factors for inequality and equality constrained models. *Journal of Statistical Software.*

Mulder, J., Hoijtink, H., and Klugkist, I. (2010). Equality and inequality constrained multivariate linear models: objective model selection using constrained posterior priors. *Journal of Statistical Planning and Inference, 140,* 887–906.

Mulder, J., Klugkist, I., van de Schoot, R., Meeus, W., Selfhout, M., and Hoijtink, H. (2009). Bayesian model selection of informative hypotheses for repeated measurements. *Journal of Mathematical Psychology, 53,* 530–546.

Ntzoufras, I. (2009). *Bayesian Model Selection Using WinBUGS.* New York: John Wiley & Sons.

O'Hagan, A. (1995). Fractional Bayes factors for model comparisons (with discussion). *Journal of the Royal Statistical Society, Series B, 57,* 99–138.

Perez, J.M. and Berger, J.O. (2002). Expected posterior prior distributions for model selection.*Biometrika,89,* 491–511.

Popper, K. (1959). *The Logic of Scientific Discovery.* London: Hutchinson.

Raftery, A.E. (1995). Bayesian model selection in social research. *Sociological Methodology, 25,* 111–163.

Robert, C.P. and Casella, G. (2004). *Monte Carlo Statistical Methods.* New York: Springer.

Robertson, T., Wright, F.T., and Dykstra, R.L. (1988). *Order Restricted Statistical Inference.* New York: Wiley.

Romeijn, J. W., van de Schoot, R., and Hoijtink, H. (In Press). One size does not fit all: proposal for a prior-adapted BIC. In: D. Dieks, W. Gonzales, S. Hartmann, F. Stadler, T. Uebel, and M. Weber (Eds.), *Probabilities, Laws, and Structures.* Berlin: Springer.

Rosenthal, L., Rosnow, R.L., and Rubin, D.B. (2000). *Contrasts and Effect Sizes in Behavioral Research: A Correlational Approach.* Cambridge: Cambridge University Press.

Rosnow, R. and Rosenthal, R. (1989). Statistical procedures and the justification of knowledge in psychological science. *American Psychologist, 44,* 1276–1284.

Roverato, A. and Consonni, G. (2004). Compatible prior distributions for DAG models. *Journal of the Royal Statistical Society, Series B, 66*, 47–62.

Royall, R. (1997). *Statistical Evidence. A Likelihood Paradigm.* New York: Chapmann & Hall/CRC.

Rubin, D.B. (1984). Bayesianly justifiable and relevant frequency calculations for the applied statistician. *The Annals of Statistics, 12*, 1151–1172.

Rutherford, A. (2001). *Introducing ANOVA and ANCOVA.* London: SAGE.

Ryan, T.P. (2007). *Modern Experimental Design.* New York: Wiley.

Schervish, M.J. (1996). *P* Values: What they are and what they are not. *The American Statistician, 50*, 203–206.

Sellke, T., Bayarri, M.J., and Berger, J.O. (2001). Calibration of *p* values for testing precise null hypotheses. *The American Statistician, 55*, 62–71.

Severini, T.A. (2000). *Likelihood Methods in Statistics.* Oxford: Oxford University Press.

Sijtsma, K. and Molenaar, I.W. (2002). *Introduction to Nonparametric Item Response Theory.* Thousand Oaks, CA: SAGE.

Silvapulle, M.J. and Sen, P.K. (2004). *Constrained Statistical Inference: Order, Inequality and Shape Constraints.* London: Wiley.

Sober, E. (2002). Bayesianism. Its scope and limits. In: R. Swinburn (Ed.), *Bayes' Theorem,* Proceedings of the British Academy Press, 113, 21–38.

Spiegelhalter, D.J., Best, N.G. Carlin, B.P., and van der Linde, A. (2002). Bayesian measures of model complexity and fit. *Journal of Royal Statistical Society, series B, 64*, 583–639.

Stephens, M. (2000). Dealing with label switching in mixture models. *Journal of the Royal Statistical Society, Series B, 62*, 795–809.

Van de Schoot, R., Dekovic, M., and Hoijtink, H. (2010). Testing inequality constrained hypotheses in SEM models. *Structural Equation Modelling, 17*, 443–463.

Van de Schoot, R., Hoijtink, H., Brugman, D., and Romeijn, J.W. (Unpublished). A prior predictive loss function for the evaluation of inequality constrained hypotheses.

Van de Schoot, R., Hoijtink, H., and Doosje, S. (2009). Rechtstreeks Verwachtingen Evalueren of de Nul Hypothese Toetsen? Nul Hypothese Toetsing versus Bayesiaanse Model Selectie [Directly Evaluating Expectations or Testing the Null Hypothesis: Null Hypothesis Testing versus Bayesian Model Selection.]. *De Psycholoog 4*, 196–203.

Van de Schoot, R., Hoijtink, H., Mulder, J., van Aken, M.A.G., Orobio de Castro, B., Meeus, W., and Romeijn, J.W. (2011). Evaluating expectations about negative emotional states of aggressive boys using Bayesian model selection. *Developmental Psychology, 47*, 203–212.

Van de Schoot, R., Hoijtink, H., and Romeijn, J.W. (2011). Moving beyond traditional null hypothesis testing: Evaluating expectations directly. *Frontiers in Psychology, 2*, Article 24.

Van de Schoot, R., Mulder, J., Hoijtink, H., van Aken, M.A.G., Dubas, J.S., Orobio de Castro, B., Meeus, W. and Romeijn, J.W. (In Press). Psychological Functioning, Personality and Support from family: An Introduction to Bayesian Model Selection. *European Journal of Developmental Psychology*.

Van de Schoot, R. and Strohmeier, D. (2011). Testing informative hypotheses in SEM increases power: An illustration contrasting classical hypothesis testing with a parametric bootstrap approach. *International Journal of Behavioral Development, 35,* 180–190.

Van de Schoot, R. and Wong, T. (In Press). Do antisocial young adults have a high or a low level of self-concept? *Self and Identity.*

Van Deun, K., Hoijtink, H., Thorrez, L., van Lommel, L., Schuit, F., and van Mechelen, I. (2009). Testing the hypothesis of tissue-selectivity: The Intersection-Union Test and a Bayesian approach. *Bioinformatics, 25,* 2588–2594.

Van Rossum, M., van de Schoot, R., and Hoijtink, H. (In Press). Inequality constrained hypotheses for ANOVA. *Methodology.*

Van Well, S., Kolk, A.M., and Klugkist, I.G. (2008). Effects of sex, gender role identification, and gender relevance of two types of stressors on cardiovascular and subjective responses: Sex and gender match/mismatch effects. *Behavior Modification, 32,* 427–449.

Van Wesel, F., Alisic, E., and Boeije, H. (Unpublished). On the origin of hypotheses: Eliciting meta-analysis, meta-synthese and experts.

Van Wesel, F., Hoijtink, H., and Klugkist, I. (In Press). Choosing priors for constrained analysis of variance. Methods based on training data. *Scandinavian Journal of Statistics.*

Van Wesel, F., Klugkist, I., and Hoijtink, H. (Unpublished). Formulating and evaluating interaction effects.

Verdinelli, I. and Wasserman, L. (1995). Computing Bayes factors using a generalization of the Savage–Dickey density ratio. *Journal of the American Statistical Association, 90,* 614–618.

Wagenmakers, E.-J. (2007). A practical solution to the pervasive problems of p values. *Psychonomic Bulletin and Review, 14,* 779–804.

Wainer, H. (1999). One cheer for null hypothesis significance testing. *Psychological Methods, 4,* 212–213.

Weiss, D.J. (2006). *Analysis of Variance and Functional Measurement: A Practical Guide.* Oxford: Oxford University Press.

Wellek, S. (2003). *Testing Statistical Hypotheses of Equivalence.* New York: Chapmann & Hall/CRC.

Wetzels, R., Grasman, R.P.P.P., and Wagenmakers, E.-J. (2010). An encompassing prior generalization of the Savage–Dickey density ratio. *Computational Statistics and Data Analysis, 54,* 2094–2102.

Zeger, S.L. and Karim, M.R. (1991). Generalized linear models with random effects: A Gibbs sampling approach. *Journal of the American Statistical Association, 86,* 79–86.

Index